Management of
Pests and Pesticides

T0139283

Westview Studies in Insect Biology
Michael D. Breed, Series Editor

About the Book and Editors

The contributors discuss pesticide production, distribution, and use; the problems caused by pests and pesticides; and the role of integrated pest management in minimizing these problems. Part I describes research at the international and regional level, and Part II addresses individual farmers' pesticide usage, their perceptions of pests and pesticides, and the relationship of the farmers' practices and attitudes to the development of integrated pest-management systems. The book concludes with a summary of research achievements and a discussion of future prospects.

Joyce Tait is senior lecturer in the Systems Department, Faculty of Technology, at the Open University in the United Kingdom and coordinator of the Research Programme on Environmental Issues of the U.K. Economic and Social Research Council. **Banpot Napompeth** is assistant professor in the Department of Entomology at Kasetsart University in Bangkok and director of the National Biological Control Research Center in Thailand.

Management of Pests and Pesticides

Farmers' Perceptions and Practices

edited by
Joyce Tait and Banpot Napompeth

Routledge
Taylor & Francis Group

LONDON AND NEW YORK

First published 1987 by Westview Press

Published 2018 by Routledge
52 Vanderbilt Avenue, New York, NY 10017
2 Park Square, Milton Park, Abingdon, Oxon OX14 4RN

Routledge is an imprint of the Taylor & Francis Group, an informa business

Library of Congress Cataloging-in-Publication Data
Management of pests and pesticides.
   (Westview studies in insect biology)
   Includes index.
   1. Agricultural pests—Integrated control. 2. Pesti-
cides. 3. Farmers. I. Tait, Joyce. II. Banpot Napompeth.
III. Series.
SP950.M346 1987   632'.9   86-1627

ISBN 13: 978-0-367-00872-7 (hbk)
ISBN 13: 978-0-367-15859-0 (pbk)

Contents

PART II
PEST MANAGEMENT DECISION MAKING
AT THE COMMUNITY LEVEL

List Of Authors

M. AHMAD, Ittehad Pesticides, P.O. Box 886, Lahore, Pakistan.

O.D. ATTEH, Department of Geography, University of Ilorin, Ilorin, Nigeria.

J.P. BANDONG, Department of Entomology, International Rice Research Institute, P.O. Box 33, Manila, Philippines.

R. BLACK, Plant Protection Section, Central Farm, Cayo District, Belize, Central America.

S. CARR, Systems Group, Faculty of Technology, The Open University, Walton Hall, Milton Keynes, MK7 6AA, U.K.

W.T. CONELLY, Winrock International, Institute for Agricultural Development, Small Ruminant CRSP, P.O. Box 252, Maseno, Kenya.

P. DEARDEN, Department of Geography, University of Victoria, P.O. Box 1700, Victoria, British Columbia V8W 2Y2, Canada.

J.P. EVENSON, Thai-Australian Prince of Songkla University Project, P.O. Box 123, Hat Yai 90110, Thailand.

I. FAGOONEE, School of Agriculture, The University of Mauritius, Reduit, Mauritius.

A.C. GOLDMAN, International Institute of Tropical Agriculture, Oyo Road, PMB 5320, Ibadan, Nigeria.

P. GUILLET, Overseas Scientific and Technical Research Office, 24 rue Bayard, 75008 Paris, France.

K.L. HEONG, Central Research Laboratories, Malaysian Agricultural Research and Development Institute, Serdang, P.O. Box 12301, Kuala Lumpur 01-02, Malaysia.

N.K. HO, Muda Agricultural Development Authority Headquarters, Ampang Jajar, Alor Setar, Kedah, Malaysia.

M.Y. HUSSEIN, Department of Plant Protection, University Pertanian Malaysia, 43400 Serdang, Selangor, Malaysia.

N. JONGLAEKHA, Faculty of Agriculture, Chiang Mai University, Chiang Mai 50002, Thailand.

P. KENMORE, Food and Agriculture Organization-Intercountry Integrated Pest Control Rice Programme, P.O. Box 1864, Manila, Philippines.

A.B. LANE, Systems Group, Faculty of Technology, The Open University, Walton Hall, Milton Keynes, MK7 6AA, U.K.

G.S. LIM, Pest Science Unit, Central Research Laboratories, Malaysian Agricultural Research and Development Institute, Serdang, P.O. Box

12301, Kuala Lumpur 01-02, Malaysia.

J.A. LITSINGER, Department of Entomology, International Rice Research Institute, P.O. Box 933, Manila, Philippines.

C.P. MEDINA, Program on Environmental Science and Management, University of the Philippines at Los Banos, College, Laguna 3720, Philippines.

V.C. MOHAN, Sahabat Alam Malaysia, 37 Lorong Birch, 10250 Pulau Pinang, Malaysia.

J. MOUCHET, Office de la Recherche Scientifique et Technique, Outre-Mer, rue d'Orsel 59, F-75018 Paris, France.

B. NAPOMPETH, National Biological Control Research Center, Kasetsart University, P.O. Box 9-52, Bangkok 10900, Thailand.

A.S. OBORDO, Department of Social Sciences, College of Arts and Sciences, University of the Philippines at Los Banos, College, Laguna, Philippines.

S.H. ONG, Pest Science Unit, Central Research Laboratories, Mardi, Serdang, P.O. Box 12301, Kuala Lumpur 01-02, Malaysia.

M.M. SALAC, Bureau of Plant Industry, San Andres, Manila, Philippines.

V.PB. SAMONTE, Department of Social Sciences, College of Arts and Sciences, University of the Philippines at Los Banos, College, Laguna, Philippines.

A.C. SANTIAGO, Food and Agriculture Organization-Intercountry Integrated Pest Control Rice Programme, P.O. Box 1864, Manila, Philippines.

C.B.S.R. SHARMA, School of Energy, Environment and Natural Resources, Madurai Kamaraj University, Madurai 625021, India.

E.J. TAIT, Systems Group, Faculty of Technology, The Open University, Walton Hall, Milton Keynes, MK7 6AA, U.K.

V. THANORMTHIN, Entomology Section, Chiang Mai Field Crops Research Centre, Maejo, Amphoe San Sai, Chiang Mai, Thailand 50210.

E.M. TUKAHIRWA, Department of Zoology, Makerere University, P.O. Box 7062, Kampala, Uganda.

H. WAIBEL, Thai-German Plant Protection Programme, Department of Agricultural Extension, P.O. Box 9-100, Bangkok 10900, Thailand.

A. YOUDEOWEI, Department of Agricultural Biology, University of Ibadan, Ibadan, Nigeria.

I.H. ZAIDI, Department of Geography, University of Karachi, Karachi-32, Pakistan.

Acknowledgments

This book would have been impossible without the help and encouragement of a great many people and organizations.

Joyce Tait began editorial work on the papers during a Rockefeller Study Fellowship at the Bellagio Conference and Study Center, Lake Como, Italy, in 1985. A grant from the US General Services Foundation, administered via Clark University in Worcester, Massachusetts, has funded several research projects and the attendance of some delegates at the Chiang Mai conference of the network on Perception and Management of Pests and Pesticides (PMPP) and has also supported the administration of the network for the past three years. Financial support has also been provided by the Food and Agricultural Organization, USAID Consortium for International Crop Protection, UNESCO Man and the Biosphere Program, the United Nations Environment Program and the Open University, UK.

Particular thanks are due to Professors Robert Kates of Brown University and Richard Ford of Clark University for their long-term support and encouragement of the PMPP network. Members of the network are too numerous to be mentioned by name, but we would like to express our gratitude to them all for their help and co-operation over the years.

Finally, and most important, Julie Lane, administrative assistant to PMPP, has managed the network, answered correspondence and generally kept it in good shape over the past three years. She helped to organize the Chiang Mai conference and without her tireless efforts in the preparation of the camera-ready copy, this volume would not have been possible. She has been skilfully supported by the other members of the production team at the Open University including Coral Long and Angela Walters (word processing), John Wilson (computer systems) and Baz East (design). Their combined efforts enabled us to master, in the end, the frustrating problems of partially compatible word processing systems and text transfer from one computer to another.

Joyce Tait
Banpot Napompeth

1
Introduction: The Management Of Pests And Pesticides

E.J. Tait and B. Napompeth

The widespread and still-expanding use of pesticides in developed and developing countries has made an important contribution to increased crop yields and improved human health. However, the hazard which pests present to crops has not been markedly diminished and in some cases it has been exacerbated. In addition, governments have had to set up regulatory systems to control the environmental and health risks of pesticides and there is also a need for internationally administered controls. There is therefore an important and interesting reciprocal relationship between the management of pests, which constitute a natural hazard to people and to crops, reflected in many cases in financial risks, and the management of pesticides, with their attendant technological risks to their users and to society in general.

On the whole, pesticide regulators seem optimistic about the compliance of those they regulate, hence the much-repeated phrase, "Pesticides are safe when used as recommended". However, research workers with direct experience in farming communities and in the agrochemical industry have long been aware that the real world is not as tidy as the regulators assume. Practical necessity, greed, the callousness of employers about the health of their workers, and ignorance of potential side-effects on people and the environment, all reduce the safety of pesticides to varying degrees. A better understanding of the *actual* behavior of those who manufacture, distribute and use pesticides, or who give advice on their use, and of the motivations underlying such behavior, is an essential prerequisite for more effective control of pests and pesticides.

Avoiding the unnecessary use of pesticides is just as important in protecting people and the environment as other forms of regulation, and provisions of this nature have been incorporated in the agricultural policies of many governments and of the Food and Agriculture Organization. However, implementing such a policy requires as a first step, educating the farming community to recognize and assess pest problems on their crops, and then training them in integrated pest management. Even in developed countries, progress in this direction is not dramatic and the pesticide treadmill, whereby the use of pesticides on crops shows a continually increasing trend, is firmly established. Reversing the direction of this treadmill will mean replacing pesticides, a technical product whose price in the market place is currently declining, with knowledge, ideas and sometimes a change of attitude on the part of the pesticide user. This transition will not be achieved without a much better understanding than we currently possess of the attitudes, motivation and behavior of farmers in relation to crop protection.

There are many aspects of the management of pests and pesticides that can be tackled successfully within the confines of traditional research disciplines, such as organic chemistry, entomology or plant pathology. However, there is also an important place for research conducted on a more holistic basis, focusing on the links between scientific and commercial research and development, the economics of the market place, government policies and regulatory instruments and the attitudes and behavior of pest managers at all levels. This is necessary for a full understanding of the influences and constraints on the production, distribution and use of pesticides, and hence for the development of improved systems for their control. Concern about the range of inter-connected problems has prompted the setting up of an international research and development network on Perception and Management of Pests and Pesticides (PMPP).

PERCEPTION AND MANAGEMENT OF PESTS AND PESTICIDES (PMPP): AN INTERNATIONAL RESEARCH AND DEVELOPMENT NETWORK

The collection of papers in this volume has a history that goes back to a meeting of the International Geographical Union Working Group on Environmental Perception in Ibadan, Nigeria, in 1978. At this meeting, some of the discussion focused on the complementary nature of the hazards presented by pests and by pesticides, and on the importance of the hazard perceptions of the regulators and users of pesticides. The advantages of adopting a multi-disciplinary approach were well understood by those at the meeting, most of whom were geographers, but this area of research was not recognized as legitimate by funding bodies and there was no readily available outlet for publications. The small numbers of people working on such topics were generally unaware of one anothers' existence and so were unable to begin to build up a coherent research tradition.

Plans were made to set up an international network to bring together natural and social scientists with the aims of fostering cross-disciplinary and international collaboration in research and development on the production, distribution and use of pesticides, legitimizing such research, and encouraging the long-term development of appropriate research methods. The network was formally inaugurated at a meeting of geographers, social scientists and biologists in Clare College, Cambridge, UK in July 1979, with major contributions from Professors Robert Kates, Ray Smith and Ian Burton, and the editors of this book.

A preliminary selection of research themes was discussed at this meeting with the long-term aim of achieving ecologically sound, economically gainful and socially acceptable strategies for pest and pesticide management. These themes were developed over the next year culminating in a research plan which was adopted at a second workshop held in Clark University, Worcester, Mass., USA in October, 1980. Four major areas of research were proposed, as outlined below.

National Profiles of Pesticide and Pest Management Practices

In industrialized countries, there is already a substantial literature on pest problems, the use of pesticides and their regulation. However, the amount of independent research is not sufficient to ensure that deviations from optimal pest management practices will be detected, or that the pesticide control systems set up in these countries will be implemented effectively. For most countries in the developing world, even rudimentary data of this nature have not been assembled.

This component of the program planned a series of studies in developed and developing countries to provide a basis for systematic comparisons among nations and to indicate problems requiring policy consideration at the national and international levels. These would focus on: major crops, pests and control measures; pesticide laws and regulations and government involvement; pest management research and development in agriculture and public health; environmental and human health hazards associated with pest management; other pesticide uses and management problems.

Pest Managers and Pesticide Users

Farmers, farm managers and workers make most of the specific decisions about agricultural pesticide use and pest control practices. Knowledge of their relevant perceptions and related behaviors is therefore essential if efficient pest management is to be promoted successfully and appropriate reactions made to pest and pesticide hazards.

This component of the program included surveys, dealing with agriculture and public health, from a wide range of countries to provide data on decision makers' perceptions of pests, perceived risks and benefits associated with different methods of control (including natural, biological and chemical methods), details of actual control methods used, sources of information and materials related to pest control.

Detailed case histories of attempts to develop alternative pest management systems were proposed, to help avoid possible pitfalls in making plans for similar systems in other regions and for other crops. Integrated pest management (IPM) was seen as the main alternative to total reliance on chemicals, with the attendant problems of environmental contamination, human toxicity and pest resistance.

International Flows of Pesticides

The export of pesticides and pesticide manufacturing processes was becoming a cause for concern, even in 1979, due to: the fear of adverse effects on local ecosystems; concern for the disruption of local agricultural systems; hazards to users, to people living near pesticide production facilities or in areas where pesticides are heavily used, and to consumers of pesticide residues in foods.

Plans were made to study the flow of processes, products, information, risk control measures and hazardous residues between industrialized and developing countries, concentrating on four broad areas: trends in international trade in pesticides; the role of pesticides in technology transfer; regulatory processes affecting international trade; the control of pesticide residues on food traded internationally.

In addition to providing intrinsically valuable information, the outcome of this research was seen as a framework to integrate and compare findings from site-specific studies and national profiles.

Supporting Activities

The secretariat of the PMPP network was located first at the Social and Political Sciences Department of Cambridge University and then at the Systems Group in the Technology Faculty of the Open University, UK. Most of the limited funding available to the program has gone to support the travel of developing country participants to conferences. Obtaining funding for research has been the responsibility of the individuals concerned. The effectiveness of the secretariat has been considerably improved in recent years with the receipt of a grant from the US General Services Foundation, which allowed us to employ a part-time administrative assistant. This also enabled us to complete the production of a computer-based bibliography of relevant research (Tait and Lane, 1985), and it is currently funding a series of research projects on pest management of direct practical relevance to small farmers in developing countries.

As membership of the network increased to its present level of approximately 220, it became clear that many people, from a wide variety of organizations involved in crop protection and public health pest control, in developed and developing countries, shared our concerns and supported our aims. A set of guidelines drawn up by participants at the Clark University meeting (Tait, 1981a) has been widely disseminated to encourage maximum comparability among research projects carried out in very different environments. This was seen as particularly important where the people involved came from a range of different academic backgrounds in the biological and social sciences, and also from non-academic backgrounds in administration, agricultural extension, industry and non-government organizations.

Published Research Results and Special Reports

The first research results based on this plan were presented at the third PMPP meeting, held at the International Center for Insect Physiology and Ecology (ICIPE) conference center at Duduville, near Nairobi, Kenya. A selection of these papers has been published in a special issue of the journal *Insect Science and its Application* (Odhiambo, 1984). This includes the following titles: 'Pest Management and Pesticide Impacts' by David Pimentel and David A. Andow; 'Government Influence on Pesticide Use in Developing Countries' by Dale G. Bottrell; 'Investigation into Pesticide Imports, Distribution and Use in Zambia with Special Emphasis on the Role of Multi-national Companies' by Tanya Abrahamse and Angela M. Brunt; two papers by I. Fagoonee on 'Pests, Pesticides, Pesticide Legislation and Management in Mauritius' and 'Pertinent Aspects of Pesticide Usage in Mauritius'; 'Pesticides and their Utilization - A Profile for Uganda' by E.M. Tukahirwa; 'Farmers' Perception and Management of Pest Hazard - A Pilot Study of a Punjabi Village in Lower Indus Region' by Iqtidar H. Zaidi; Nigerian Farmers' Perception of Pests and Pesticides' by Oluwayomi D. Atteh; 'Pest Control Practices of Rice Farmers in Tanjong Karang, Malaysia' by K.L. Heong; and 'Problems of Vector-Borne Diseases and Irrigation Projects' by M.W. Service.

Other papers produced under the program cover a range of relevant topics. National profiles have been written for Thailand (Napompeth, 1981) and Lesotho (Turner and Zinyowera, 1979). International flows of pesticides and national regulatory systems have been discussed by Napompeth (1979) and Tait (1981b). Research on pest and pesticide perceptions of British farmers has been described by Tait (1983).

FOURTH INTERNATIONAL PMPP MEETING, CHIANG MAI, THAILAND

The fourth PMPP Meeting was organized in Chiang Mai, Thailand in January, 1985 by Dr. Banpot Napompeth, Director of the National Biological Control Research Center, Kasetsart University, as the culmination of the research plan drawn up in 1980 at Clark University.

The conference papers are presented here in two main groups, dealing with research at the 'macro' and 'micro' levels. Part I covers the management of pests and pesticides from a regional, national or international perspective. It therefore deals mainly with the topics outlined in the Research Plan under the headings 'International Flow of Pesticides' and 'National Profiles of Pesticide and Pest Management Practices', including problems created by pesticides; the measurement of pesticide production and use and the effectiveness of controls on these; the design and implementation of pest management systems at the regional level; and the state provision of education, extension and pest control materials.

The papers in Part II come under the heading of 'Pest Managers and Pesticide Users' in the Research Plan. They present the results of research on decision making at the farm level in many different countries, studying individual farmers' pesticide usage, their perceptions of pests and pesticides, extension and education needs and the relationship of these to the development of integrated pest management systems.

The final chapter summarizes the conference discussions, outlining some of the policy implications of the research already carried out and indicating promising areas for further research.

Some of the research described in these papers was carried out under extremely difficult circumstances. Also, given the varied nature of PMPP membership, some papers are not, strictly speaking, reports of academic research; they reflect the experience of their authors in a relevant aspect of the management of pests and pesticides, and are therefore a useful information resource in themselves.

Both pests and pesticides present serious potential hazards to people and to the environment and our lack of knowledge and understanding of how they are actually managed in the real world should be a cause for concern. The papers presented here cannot do more than begin to indicate how we could tackle such problems. However, viewed from another perspective they constitute a major achievement - working with minimal resources and sometimes in the face of serious difficulties, a group of enthusiastic research workers has contributed significantly to laying the foundations for an important new area of research.

REFERENCES

Odhiambo, T.R. (ed.) (1984). Special issue: perception and management of pests and pesticides. *Insect Science and its Application*, 5(3): 139-231.

Napompeth, B. (1979). Socio-economic aspects of pest management in Thailand. Paper presented to Seminar on the Perception of Pests and Pesticides as Environmental Hazards, Cambridge, UK, June 1979.

Napompeth, B. (1981). Thailand national profile on pest management and related problems. Special Publication 4, National Biological Control Research Center, Kasetsart University, Bangkok, Thailand.

Tait, E.J. (ed.) (1981a). Perception and management of pests and pesticides: guidelines for research. Working Paper EPR-8, Publications and Information, Institute for Environmental Studies, University of Toronto, Toronto, Canada, M55 1A4.

Tait, E.J. (1981b). The flow of pesticides: industrial and farming perspectives. In *Progress in Resource Management and Environmental Planning*, Vol. 3, eds. T.O. Riordan and R.K. Turner, pp 219-250. Chichester: John Wiley and Sons.

Tait, E.J. (1983). Pest control decision making on brassica crops. In *Advances in Applied Biology*, Vol VIII, ed. T.H. Coaker, pp 122-188 London: Academic Press.

Tait, J. and Lane, J. (1985) Perception and management of pests and pesticides: reference bibliography. Biosystems Research Group, Technology Faculty Open University, Milton Keynes, UK MK7 6AA.

Turner, S.D. and Zinyowera, M.C. (1979). National profile of pest management: Lesotho Paper presented to Seminar on the Perception of Pests and Pesticides as Environmental Hazards, Cambridge UK, June 1979.

PART I
PEST AND PESTICIDE MANAGEMENT AT THE INTERNATIONAL, NATIONAL AND REGIONAL LEVELS

2

Environmental Problems Of Pesticide Usage In Malaysian Rice Fields — Perceptions And Future Considerations

Lim Guan-Soon and Ong Seng-Hock

INTRODUCTION

In the past two decades the Malaysian government has given high priority to agricultural development. Irrigation has enabled the planting of two crops per year or three crops in two years, and extensive cultivation of short-term and high-response varieties (HRV) covering more than 70% of the total paddy area.

Traditional methods of cultivating local low-yielding rice varieties are characterized by minimal pesticide use, thus allowing a significant degree of natural control of pests and resistance of the cultivars to infestations. The fauna of such paddy fields is very diverse. Under the new system, the technological package (HRV of paddy, abundant use of fertilizers and pesticides, and irrigation), along with price-support policies, has resulted in a rapid increase in crop yields.

The recommended list of pesticides for use on rice includes: the insecticides dieldrin and DDT (to be used only where other pesticides are not effective), acephate, butylphenyl carbamate, carbaryl, carbofuran, dimethoate, diazinon, endosulfan, fenthion, fenitrothion, gamma HCH, isoprocarb, malathion, phenthoate, propoxur, quinalphos; the fungicides benomyl, blasticidin-S, edifenphos, isoprothiolane, mancozeb, thiram, tricyclazole; the herbicides 2,4-D, MCPA, paraquat and propanil; the rodenticides calcium cyanide, zinc phosphide, coumatetralyl and warfarin (Ministry of Agriculture, 1981a; Supaad et al., 1982).

Surveys have shown that 89%, 83% and 45% of the farmers in Tanjong Karang, Krian and Muda respectively, used insecticides. In Tanjong Karang, 41%, and in Krian 26%, sprayed at least three times. Prophylactic spraying, although declining, is still practised by 16% of farmers in Tanjong Karang and 26% in Krian (Department of Agriculture, 1982; Ho, 1982). Herbicide, usually paraquat or 2,4-D, is used commonly only in Tanjong Karang and Sg. Manik. In Kelantan, Krian and Pahang, this practice is on a small scale.

Crop losses are so high in some places that there are good arguments for using more pesticides, rather than less. However, there may be associated effects on human welfare, including damage to health, economic and resource deprivation (such as a decline in fish harvests), interference with control of pests by beneficial insects, an effect on the perceived 'quality of life', and physical damage to crops and the environment. This paper aims to examine these effects and to assess their relative importance, taking account of developing trends and the general perception of the problems, and examining efforts to overcome them.

ENVIRONMENTAL EFFECTS OF PESTICIDES

Acute human poisoning from pesticides is easier to detect than chronic effects. During an outbreak of the brown planthopper in Tanjong Karang in 1977, carbofuran granules (3%) were widely broadcast. At least 17 farmers were hospitalized after inhaling fine dust, the carbofuran content of which was much higher than in the intact granules. In another poisoning incident in June 1979 in Kedah (Zain, 1979; Sahabat Alam Malaysia, 1981), within one week one farmer died and over 30 were admitted to hospital with suspected pesticide poisoning. Many of the poor farmers are not properly trained to use pesticides, and at least 12 of them were using motorised mistblowers for the first time. A less serious effect of pesticides was reported in early 1980 when paddy farmers complained of swelling in their legs and body rashes (Sahabat Alam Malaysia, 1982). They were found to be using pesticides indiscriminately.

Farmers who use pesticides are often illiterate. They rarely follow instructions and often rely on shop owners, salesmen and other farmers for recommendations on the use of pesticides. Farmers rarely use any protective clothing and are ignorant of the safety procedures in handling chemicals.

The long-term consequences of chronic exposure to low concentrations of pesticides may be more serious than those of acute pollution and the cause may be more difficult to detect. Also, some breakdown products may be more toxic than the pesticides themselves. Extra caution should be exerted for chemicals with long half-lives.

Analysis of organochlorine levels in paddy and soil from the Tanjong Karang area, where they have been widely used, detected low level residues of HCH and endosulfan only: for alpha and gamma isomers of HCH in paddy field water, 0.001 ppm or less; for HCH in paddy field soils, the beta isomer was generally undetectable, and alpha and gamma isomers were generally below 0.05 ppm, but six results out of 42 ranged up to 1.06 ppm; endosulfan residues in paddy soils were generally undetectable or present only in traces, but six samples out of 42 ranged up to 0.7 ppm; endosulfan residues in water containing dead fish ranged from 0.001 to 0.03 ppm for the alpha and beta isomers and were undetectable or present only in traces for the sulfate.

The low residue levels are probably due to high temperature, and humidity, and anaerobic soil conditions (Siddaramappa and Sethunathan, 1975; MacRae et al., 1967). Microbes have been implicated in the breakdown of HCH in flooded soils (Mathur and Saha, 1975). The presence of alpha-HCH in all the samples indicates that this isomer is persistent and perhaps cumulative in the rice environment. Cheaper HCH formulations containing mixed isomers are often used by farmers, probably explaining the consistent presence of alpha HCH in the samples. Beta-HCH was found only in soils. Although absent from water, endosulfan was found in soils where it may be more persistent. Although residue levels were generally low, their presence may still not be desirable environmentally, as they may enter the tissues of the aquatic fauna and accumulate in paddy field fish, which are an important food source.

Among paddy farmers, chronic exposure to pesticides occurs through routine spray operations. Mixing pesticides with bare hands can also lead to harmful effects, but these have not been monitored in Malaysia. Such a study should be initiated as soon as possible.

Children, women and old people who are not involved in spray operations may be exposed to pesticides in contaminated irrigation water, which is used for domestic activities where tap water is not easily available. Often, the same canal water is used for cleaning spray equipment or even disposal of empty pesticide containers.

Chronic exposures to pesticides may also occur via residues in food and drinking water. However, up-take of pesticide residues from field treatment of the crop is likely to be negligible as the harvested paddy is milled before consumption. Only traces of HCH were found on paddy grains taken from the farmers' fields, suggesting that it is not easily translocated to the paddy grains. High residues have been detected in rice (Table 2.1) as a result of storage treatments against pests.

Up-take of residues through the consumption of paddy field fish has long been a concern because of the large quantity harvested and consumed locally. Only HCH residues were detected in the fish Sepat Ronggeng (*Trichogaster trichopterus*) collected live from Tanjong Karang in 1982. Ten samples, each of three to five fish, were collected and each was replicated twice. The alpha HCH isomer ranged from 0.018 ± 0.026 ppm to 0.058 ± 0.027 ppm and the gamma isomer from 0.010 ± 0.021 ppm to 0.100 ± 0.029 ppm. Particularly high residues occurred in fish killed by HCH under experimental conditions (Table 2.2).

Endosulfan residues have also been found in dead and dying fish (Table 2.3). Two samples, each of 15 fish, dying as a result of endosulfan poisoning in the field in Tanjong Karang in 1982, were found to contain the following endosulfan residues (ppm): alpha isomer, 5.13 ± 1.42 and 4.19 ± 1.03; beta isomer, 1.70 ± 0.34 and 1.39 ± 0.26; sulfate, 0.46 ± 0.09 and 0.42 ± 0.02.

TABLE 2.1

HCH residue levels in rice grains from warehouses and milling plants (Lee & Ong, 1982)

Location	HCH residue (ppm ± standard error)	Range (ppm)
Warehouses		
Site 1	9.91 ±11.31	1.17—50.0
Site 2	0.82 ± 0.75	0.02—3.35
Site 3	0.45 ± 0.43	0.05—1.31
Site 4	0.50 ± 0.39	0.09—1.48
Milling plants		
Sungei Besar	0.04 ±0.008	0.02—0.04
Sekinchan	0.01 ±0.005	0.007—0.02
Ulu Thiram	0.03 ±0.004	0.02—0.04

TABLE 2.2

HCH residues in dead fish resulting from poisoning in the laboratory by HCH application at different concentrations

Concentration of HCH[†] (ppm)	HCH residues (ppm)	
	gamma	alpha
0.5	76.92 ±15.32	87.21 ± 7.10
1.0	61.54 ±24.86	81.04 ±10.15
2.0	284.23 ±35.91	177.59 ±17.46
3.0	184.62 ± 8.20	194.83 ±10.63
4.0	192.31 ±15.78	108.62 ±22.95
5.0	264.15 ±22.95	336.21 ±21.11
6.0	254.32 ±13.18	120.42 ±22.97
7.0	265.38 ±21.47	139.66 ±18.67

† Each residue analysis was replicated 3 times, with 6-20 dead fish per concentration.

Carbofuran, commonly used in Malaysia for paddy stemborer and brown planthopper control, was found from studies in the Philippines to have little potential for bioaccumulation. The limited evidence available suggests that fish exposed to carbofuran is safe for human consumption.

Few such studies have been done for the wide range of pesticides used in paddy fields. In the U.S., biomagnification of DDT occurred in Lake Michigan where levels ranged from 0.000002 ppm in the water to 10 ppm or more in fish (Metcalf, 1975). Accumulation of other persistent

TABLE 2.3
Endosulfan residues in dead fish resulting from poisoning by endosulfan application at different concentrations

Concentration of endosulfan (ppm)†	Endosulfan residues (ppm)		
	alpha	beta	sulphate
0.00375	0.232 ±0.128	0.162 ±0.058	0.261 ±0.089
0.0075	0.883 ±0.170	0.517 ±0.096	0.725 ±0.259
0.01	0.770 ±0.280	0.469 ±0.213	0.551 ±0.243
0.015	1.204 ±0.177	0.763 ±0.298	0.681 ±0.123
0.02	1.229 ±0.110	0.639 ±0.298	0.547 ±0.043

† Each concentration was replicated 3 times, with 6-20 dead fish per concentration used for residue analysis.

TABLE 2.4
Biomagnification of HCH and endosulfan (Tanjong Karang, Malaysia, 1982)

Pesticide	Soil (A)	Fish (B)	Water (C)	Build-up (A/C)	Biomagnfication (B/C)
alpha-HCH	0.020	0.038	0.0005	44	85
gamma-HCH	0.046	0.031	0.0001	354	239
alpha-endosulfan	—	4.66	$0.0110 \pm 4.0 \times 10^{-3}$†	—	424
beta-endosulfan	—	1.54	$0.0060 \pm 1.0 \times 10^{-3}$†	—	275

† Each value is an average of 3 samples of ikan lundu (*Mystus vittatus*) at the time of dying, resulting from endosulfan poisoning in the field.

pesticides, or their breakdown products, in aquatic systems has had similar effects (Bottrell, 1979). Further evidence of biomagnification in Malaysian paddy systems has been obtained for both HCH (x85-239) and endosulfan (x275-424) (Table 2.4).

Fish harvests from paddy fields have declined considerably since 1972, by as much as 50-60% in the Krian district (Perak) and Seberang Prai (Penang), due to the destruction of fish-breeding areas, over-exploitation of fish resources and a general deterioration in the quality of water. Fish mortality is particularly evident following the application of endosulfan (Yunus & Lim, 1971). The introduction of rice varieties with shorter maturing periods is believed to have contributed significantly to fish decline through earlier draining of water, reducing the time for completion of their reproductive cycle.

Fish are an important resource for paddy farmers and this destruction will mean poorer nutrition and more poverty. To safeguard against this, it is important that pesticides used are of low toxicity to fish. The consumption of paddy field fish is not confined to the paddy farming community and those in the immediate urban vicinity. They are consumed throughout the country and are exported to Hong Kong and other ASEAN countries. It is therefore desirable to minimize the use of chemicals that may accumulate in paddy field fish.

Epidemics of fish disease in paddy fields have frequently been reported. Public opinion attributes this to pesticides, but investigations have indicated that organic load or water impurities may be the main factor involved. Laboratory studies failed to produce diseased fish when subject to exposure of sublethal concentrations of 19 pesticides for up to three months. However, when fish were kept free from pesticides in aquaria that were not scrubbed of their slime, disease symptoms developed. The infecting agent was identified as *Aeromonas* sp., a bacterium commonly found on diseased fish in the field during the epidemics. Affected fish have external lesions on the body surface and ulcers in advanced cases (Ministry of Agriculture, 1981b). The epidemic, locally known as *wabak kudis*, was first reported in paddy fields during December 1980 to February 1981, and again late in 1981, the disease being mainly confined to the states of Kelantan, Kedah, Penang (Seberang Prai), Perak (Krian and Sungai Manik) and Melaka. Relatively less pesticide is employed in these localities compared to Tanjong Karang, where only limited incidence of wabak kudis was noted, further indicating that chemical pesticides are not involved.

Frogs and snakes are also commonly observed to be affected by pesticides in paddy fields. Both species are useful predators, of insect pests and rats respectively. The frogs are particularly susceptible to pesticides, large numbers having been found dead after the use of monocrotophos and dieldrin. A number of cattle poisoning cases have been recorded in the country, although none has been traced to the use of pesticides in paddy fields. Deaths of ducks have been observed, and although the number of cases is few, there is concern about the effects of chronic exposure.

In unsprayed paddies, planted with local rice varieties, pests are maintained at low levels by the activities of natural enemies (Lim, 1974). In pesticide trials, treated plots have frequently failed to yield more harvestable grain than the untreated controls, indicating that pest numbers are too low to affect the yield. In fields treated with insecticide, the common predators of planthoppers were reduced by 6.3% to 79.3% depending on the predator species and the chemicals employed (Supaad et al., 1982). Similar effects were noted for stemborer parasites (Lim & Heong, 1977). Such disruptions can lead to target pest resurgence and secondary pest outbreaks.

The latter may have occurred when intensive and widespread pesticide applications against brown planthoppers were followed by severe dark-head stemborer attacks.

The resurgence of planthoppers following chemical sprays may not be due to disruption of natural control. Experimental evidence has indicated that pesticides can induce planthopper resurgence through modified physiological conditions on the host plant (Chelliah et al., 1980). This may explain failures in planthopper control, reported by spray teams in the Tanjong Karang Irrigation Scheme.

Farmers and government control operators often need to increase the rate and frequency of pesticide application to obtain a satisfactory response. In the long term, this chemical treadmill has proved to be self-defeating.

PERCEPTION AND APPRAISAL OF THE ENVIRONMENTAL PROBLEMS

The environmental problems arising from the use of pesticides on paddy in Malaysia are still minor in comparison to other countries (e.g. Indonesia, Japan, the Philippines). However, concern is increasing in research and extension organizations (for example the Department of Agriculture (DOA), Malaysian Agricultural Research & Development Institute (MARDI), the Ministry of Science, Technology and Environment, and universities), with the rise in pesticide usage, following escalating pest problems.

Public awareness is generated through Sahabat Alam Malaysia (SAM, Friends of the Earth) and state consumer associations, in particular the Consumer Association of Penang (CAP). Information is disseminated by these agencies through local media, particularly newspapers, newsletters, occasional publications and seminars or symposia. Despite occasional weaknesses in the presentation of information, these agencies have performed a service in increasing public awareness of environmental problems and although branded as 'alarmists' in some quarters which have suffered from their pressure, their continued existence and support is necessary to balance the activities of irresponsible elements that would otherwise exploit the environment without restraint. They have also served as effective pressure groups on authorities with the responsibility for enforcing environmental health measures.

The perception of environmental problems arising from the use of pesticides on paddy has increased, but oversimplification of the issues such as calling for blanket withdrawal of some chemicals is not likely to result in a satisfactory solution. Considering each need in its specific context is preferable. For example, apart from its piscicidal effect, endosulfan is preferable for paddy pest control because it is effective and cheap and has few undesirable side effects. Banning its use would deprive many farmers of a potentially useful material, and restricting it to areas where fish are unimportant would be more rational.

In devising measures to minimize the environmental problems associated with the use of pesticides on paddy, there is a need to appreciate their relative importance. Acute poisoning of farmers, adverse effects on fish productivity and the effects on beneficial arthropods appear most important, the latter deserving special concern as it may govern the future need for chemicals.

MEASURES TO MINIMISE THE ENVIRONMENTAL EFFECTS OF PESTICIDES

About M$160 million worth of pesticides are sold annually in Malaysia. Control is exercised by the *Pesticide Act* (1974) (Balasubramaniam et al., 1978), with the objectives of controlling the efficiency of pesticides and their toxicity to users, the general public, domestic animals, wildlife and the environment.

There are adequate provisions for regulating the use of the available pesticides, and since the initiation of pesticide registration in 1981, many hazardous chemicals have been deregistered and their use prohibited. For example, monocrotophos, once recommended for paddy stemborer control (Lim, 1972), is no longer in use, and the import of phorate has been stopped.

It is sometimes not possible to prohibit the use of all toxic pesticides. For instance molinate, which is highly toxic to the paddy field fish, is available for control of *padi burong*, a weed which has only recently become serious in direct-seeded paddies. No effective and less hazardous alternative is presently available.

Efforts are also directed at using acceptable chemicals judiciously. This embodies the adoption of integrated pest management (IPM), involving a variety of biological, physical and chemical methods, integrated into a cohesive scheme, designed to provide long-term protection (Bottrell, 1979). Chemical pesticides are used only as a last resort, the objective being to manage the pests in an economically efficient and environmentally sound manner.

Among the tactics given emphasis in integrated control for paddy pests is breeding for resistant crop varieties. From the farmer's viewpoint this is usually the most effective, easy, and economical means of controlling pests, particularly for crops such as paddy with a relatively low value.

Chemical pesticides are still an indispensible tool in paddy pest control. They are relatively cheap and adaptable for use in a wide variety of situations, including the rapid control of planthopper outbreaks (Lim et al., 1978). However, they must be used judiciously, in relation to economic threshold levels as indicated by an extensive surveillance and forecasting scheme (Ooi, 1982). This scheme has avoided unnecessary pesticide treatment in extensive paddy areas, unlike the original situation when panic at the slightest sign of pests led to blanket application. Early detection of isolated pockets of pests elicit 'spot treatment' to prevent the development of an outbreak. Large scale use of pesticides can thus be avoided.

The surveillance scheme was first proposed by Heong (1977), but made little headway until its usefulness in combating brown planthopper outbreaks was recognized by key policy makers in 1979 (Lim et al., 1981). It was implemented initially in the Tanjong Karang Irrigation Scheme (Lim et al., 1978) and has been extended to Kedah, Perlis, Kelantan, Trengganu and Seberand Perai in Penang, and Krian and Sg. Manik in Perak (Ooi, 1982). It depends on field scouting, which, although effective, is time consuming. Improvements in detection capability and further expansion of the scheme to other paddy areas should further reduce pest outbreaks, and hence the use of pesticides.

FUTURE DEVELOPMENT AND NEEDS

A prime objective of any developing country is to achieve self-sufficiency in food production and hence to save valuable foreign exchange for development. In striving towards this objective, Malaysia has embarked on a policy of agricultural modernization, particularly in paddy cultivation. It includes the introduction of new technologies and better organization of the present farming activities. The introduction of modern technologies may sometimes bring about unexpected adverse effects such as increased pest problems and, unless efficiently managed, pesticide usage may escalate to the point where it creates an environmental crisis.

The new technology of direct seeding is creating a new range of problems. This was introduced to overcome labour shortages and increase paddy yields but it has created a new, serious weed problem, padi burong, and the position of other pests has worsened. The increase in planting density, closer canopy, and more staggered planting conditions, have created a microclimate and habitat conducive to the multiplication of pest species, and many which previously were not a problem are gradually becoming more troublesome (e.g. *Nezara viridula* L. and *Scotinophara coarctata* F.).

Because direct-seeded paddies have no in-field walking space, farmers are reluctant to allow surveillance staff to enter their plots for scouting and inspection, resulting in poor pest surveillance, with initial pest build-ups escaping detection, or attacks reported too late. Ensuing major pest outbreaks could result in excessive pesticide application and adverse environmental effects.

Labour shortages in rural areas due to industrial development, have resulted in large numbers of paddy fields being left idle, and the concept of centralized farming is being considered. Merging the farms into larger components will allow more effective use of resources and also some advantages in relation to implementation of IPM, as the timing of farm operations can be better controlled. However, during a pest outbreak, pesticide will probably be used on a much larger scale, using aerial application, with an increase in environmental problems. Subsidy schemes will further increase pesticide and fertilizer use. Intended as an aid to the farmers, they encourage liberal use of these products and will present difficulties for the implementation of IPM.

The above developments generally have negative effects on the environment and only the successful implementation of IPM programs can minimize them. Many policy makers are sympathetic to IPM, as reflected in the support given by the government to the 'FAO Intercountry Integrated Pest Control Program for Rice', implemented in 1981. This has acted as a catalyst for IPM developments in Malaysia, resulting in the setting up of the National IPC Committee which plans, reviews, and implements IPM projects throughout the country (Chang et al., 1980).

The FAO intercountry program has increased training opportunities for scientists, agriculturists and technicians to help promote IPM research and extension. It has also generated greater awareness in crop protection scientists and extension personnel of the importance of social constraints on IPM. This has led to a better understanding of farmers' pest control decision making and has resulted in the development of more realistic management tactics and improvements in the education process for farmers. In the commercial sector, rapidly escalating research and development costs, increased difficulties in registration, growing concern for the environment, and a decreasing rate of new discovery of pesticides have focused attention on new frontiers such as natural products and bio-rational pesticides (pheromones, insect growth regulators and anti-juvenile hormones), which promise to deliver safe, cost effective, selective, and environmentally acceptable methods of pest control (Menn, 1980). However, there are many uncertainties still surrounding such developments. In view of this, farmers, extension workers and research personnel must keep a sense of proportion. Ecological damage from dangerous pollutants is real, and there is no guarantee that satisfactory alternatives will be available in time to avert an environmental crisis. The question is not whether conventional pesticides should continue to be used. Rather, it is how they may be used with minimum undesirable side effects and complications, in line with the IPM concept.

REFERENCES

Balasubramaniam, A., Pan, S.H., Yeoh, H.F. and Zain, M.A.R. (1978). Pesticide legislation in Malaysia. In: *Proceedings Plant Protection Conference, Kuala Lumpur 1978*, eds. L.L. Amin et al., Rubber Research Institute. Malaysia, Kuala Lumpur, pp. 242-50.

Bottrell, D.G. (1979). *Integrated Pest Management*. Council on Environmental Quality, Washington.

Chang, P.M., Lim, G.S., Supaad, M.A. and Lam, Y.M. (1980). Development of integrated pest control in rice in Malaysia. In *The National Rice Conference 1980*. MARDI, Serdang, pp. 191-201.

Chelliah, S., Fabellar, L.T. and Heinrichs, E.A. (1980). Effect of sub-lethal doses of three insecticides on the reproductive rate of the brown planthopper, *Nilaparvata lugens*, on rice. *Environmental Entomology* 9: 778-80.

Department of Agriculture (1982). Pesticide usage in rice production. Pesticide Section, Crop Protection Branch, Department of Agriculture, Kuala Lumpur, Malaysia.

Heong, K.L. (1977). Pest surveillance and monitoring system for rice cultivation in Tanjong Karang — a proposal. Rice Research Branch MARDI, Malaysia (mimeographed).

Ho, N.K. (1982). Pest problems related to direct seeding in the Muda area. Malaysian Plant Protection Society (MAPPS) Newsletter 6 (3): 4-5.

Lee, B.S. and Ong, S.H. (1982). Problems associated with pesticide use in Malaysia. Paper presented at the international symposium on pesticide use in developing countries — present and future. Kyoto, Japan. 31 Aug-Sept. 1982.

Lim, G.S. (1972). Chemical control of rice insects and diseases in Malaysia. *Japan Pesticide Information* 10: 27-36.

Lim, G.S. (1974). Potential for the biological control of rice insect pests. MARDI Report 22.

Lim, G.S. and Heong, K.L. (1977). Habitat modifications for regulating pest population of rice in Malaysia. MARDI Report 50.

Lim, G.S., Heong, K.L. and Ooi, A.C. (1981). Constraints to integrated pest control in Malaysia. In: *Conference on Future Trends of Integrated Pest Management.* International Organization of Biological Control Special Issue, Center for Overseas Pest Research, London, pp 61-6.

Lim, G.S., Ooi, A.C. and Koh, A.K. (1978). Outbreak and control of the rice brown planthopper (*Nilaparvata lugens* Stal.) in Tanjong Karang, Malaysia. In: L.L. Amin et al. (eds). *Proceedings Plant Protection Conference Kuala Lumpur 1978.* Rubber Research Institute, Malaysia, Kuala Lumpur, pp. 193-213.

MacRae, Raghu, K. and Castro, T.F. (1967). Persistence and biodegradation of four common isomers of benzene hexachloride in submerged soils. *Journal of Agriculture Food Chemistry* 15: 911-4.

Mathur, S.P. and Saha, J.G. (1975). Microbial degradation of lindane — C5144 in a flooded sandy loam soil. *Soil Science* 120: 301-7.

Menn, J.J. (1980). Contemporary frontiers in chemical pesticide research, *Journal of Agriculture Food Chemistry* 28: 2-8.

Metcalf, R.L. (1975). Insecticides in pest management. In : *Introduction to Insect Pest Management.* eds. R.L. Metcalf and W. Luckman. Wiley-Interscience, New York, pp 235-73.

Ministry of Agriculture (1981a). Syor-Syor Rawalan Tanaman 1981. Ministry of Agriculture, Kuala Lumpur, Malaysia.

Ministry of Agriculture (1981b). Annual fisheries statistics 1980. Ministry of Agriculture Malaysia.

Ooi, P.A.C. (1982). A surveillance system for rice planthoppers in Malaysia. In: International Conference on Plant Protection in the Tropics. eds. K.L. Heong et al. Malaysian Plant Protection Society, pp. 551-65.

Sahabat Alam Malaysia (1981). Pesticide problems in a developing country — a case study of Malaysia. Sahabat Alam Malaysia, Penang.

Sahabat Alam Malaysia (1982). The State of the Malaysian environment 1981-82. Sahabat Alam Malaysia, Penang.

Siddaramappa, R. and Sethunathan, N. (1975). Persistence of gamma-BHC and beta — BHC in Indian rice soils under flooded conditions. *Pesticide Science* 6: 395-403.

Supaad, M.A., Lim, B.K. and Yong, Y.C. (1982). Pesticides (insecticides) use and their specificity on rice in Peninsular Malaysia. Paper presented at: Meeting of the Organizers of the Working Group on Pesticide Use and Specificity on Rice FAO/RAPA, 23-36 Nov. 1982, Bangkok.

Yunus, A. and Lim, G.S. (1971). A problem in the use of insecticides in paddy fields in West Malaysia — a case study. *Malaysian Agricultural Journal* 48: 167-78.

Zain, M.A.R. (1979). Lapuran mengenai Kejadian terkena racun pentani-pentani Kedah/Perlis Jun, 1979. Cawangan Pemeliharaan Tanaman, Jab. Pertanian, Kuala Lumpur. (mimeographed, in Bahasa Malaysia).

3

Educating And Training Pest Managers: The Role Of Distance Teaching

A.B. Lane and E.J. Tait

INTRODUCTION

Pest management systems, both integrated and chemically based, in developed and developing countries, are becoming increasingly complex and demanding, beyond the ability and knowledge of the average farmer or farm manager. A common response to this problem is to put crop protection decision making in the hands of experts, either in government extension services or in the commercial sector. Government services are rarely funded on a sufficiently lavish basis to allow them to give regular advice to large numbers of farmers. On the other hand, commercial companies, with their need to maintain growth in sales in order to fund their research and development activities (Tait, 1981), are often prepared to take up this challenge. In Britain, where the government advisory service has been experiencing financial cuts for a number of years, a survey of arable farmers found that over 50% had a representative of a commercial pesticide distributor walking their fields once a week and making decisions on crop protection needs. It would be naive to presume that the advice given under these circumstances did not sometimes recommend the use of pesticides where none was needed (Tait, in press).

Crop protection systems that rely on off-farm expertise are bound to be vulnerable to a range of pressures. A major aim of any advisor, government or commercial, is to maintain his or her credibility, leading in both cases to a tendency to make risk averse decisions and hence, to a greater or lesser extent, to guide farmers onto a pesticide treadmill. Also, in times of crisis such as major pest outbreaks or the sudden emergence of a new pest, external experts will find it difficult to cope. A more robust and stable pest management system would be one based on an appropriately educated and trained population of farmers. Under these circumstances decision makers have a different kind of stake in the decisions and may even be able to use informed judgement and delay applying pesticides, secure in the knowledge that they can monitor the situation and take rapid action if necessary.

Two major prerequisites to achieving such a situation are: creating an awareness in the farming community of the dangers of losing control of crop protection decision making and hence creating a demand for the required training: delivering the training quickly and effectively to large numbers of people. This paper deals with the latter problem, which has been identified as a constraint on the adoption of IPM (Corbet, 1981), concentrating on the distance teaching methods developed in Britain by the Open University.

DISTANCE TEACHING METHODS

Open University students require no prior qualifications, even for entry to degree level courses (Morris, 1979). They study at home, which can be anywhere in the United Kingdom, and they are often studying part-time while in full-time employment. Distance teaching is made possible by the use of a range of media to achieve specific teaching aims within an integrated package.

Written texts are characterized by a particular style which is clear, concise, immediate and lavishly illustrated, sometimes using cartoons to convey ideas in a memorable way (Merrill, 1978; Turpin and Brown, 1979) (Figure 3.1). The text is interspersed with self-assessment questions and, where appropriate, there are cross references to other course components.

Either broadcast television or video cassettes can be used to make visual teaching points which are difficult to convey via the printed word or static pictures. Television can usually reach large numbers of people and, where video cassette recorders are not available, may be the only option. However, its teaching potential is limited compared to video cassettes, where the stop/start facility allows for more direct interaction between the student and the medium. Similar distinctions can be made between radio and audio-cassettes, important roles for these media being to 'talk students through' complex diagrams or arithmetic calculations, or, by means of recorded discussions to represent a range of opinion.

A course or teaching package usually has a life of five to eight years after which it will require updating. It is conceived, developed and maintained throughout its life by a course team. They decide on its aims and objectives, devise its structure, and produce or supervise the production of the various components. A very important element in the success of the Open University is the arrangements made to ensure that students have personal contacts with their teachers and fellow students, through part time tutors who live in their own region and who mark their regular assignments, and also through attending summer schools.

DISTANCE TEACHING PACKS IN CROP PROTECTION

The Science and Technology Updating Sector of the Open University is developing a series of teaching packages for farmers and others involved in agriculture (Morris, 1985). The first two, produced in response to the needs outlined in the introduction, have dealt with crop protection on oilseed rape and winter cereals.

Inspecting the crop for problems: what does the crop look like and is it healthy?

Inspecting the crop can take many forms, from a quick look as you pass it by in the car to doing extensive scouting or field-walking. It is up to you how much time you spend on this, as long as you know the limitations of each method.

A quick look from the edge of a crop is not very useful for two reasons. Firstly, pest and disease attack tends to be greater on the headlands making them different from the rest of the field. Secondly, because you see nothing to worry about at the edge, do not assume that there is not something the matter, or out of sight, in the middle of the crop.

You will only gain a clear picture of what is happening on your crop by walking through it, looking for signs of damage or disease and the presence of pest organisms.

Walk into the middle of the crop.

Figure 3.1   The use of a cartoon to emphasise a teaching point.

An important decision, taken early in their development, was to aim to educate students in the underlying ecological and economic principles of crop protection, in addition to training them in the tactical aspects of managing pests and diseases. This was seen as essential if the trainees were to be flexible enough to cope with unexpected or new situations and also to begin to make logical connections between different pest control actions that would allow them eventually to understand and implement integrated pest management systems. The aims of the packs are:

i)   to improve the student's basic understanding of pest and disease problems in agriculture in general and on their farms;

ii)   to teach students to identify the important pests, predators and diseases on the crop in question;

iii) to teach students how to assess levels of infestation of pests and diseases and where possible to forecast the outcome of the infestation;

iv)   to enable students to decide, on an economically sound basis, the need to treat for pests and diseases and so to avoid unnecessary use of pesticides and unexpected pest infestations;

v)   to teach the value of keeping records to help pest management decisions.

Each pack consists of two teaching texts, a video cassette, an audiocassette, field identification cards, recording charts and a study guide indicating how to use the pack, all enclosed in a rigid plastic folder. The books, *Understanding Crop Protection* (Lane, 1984a) and the crop-specific *Handbooks* (Lane, 1984b, 1985), are the core materials to which everything else relates. The standard study time for a complete pack is 20-25 hours. *Understanding Crop Protection* teaches the principles of pest and disease ecology, economic aspects of decision making and practical aspects of crop protection, in a general way, not related to any particular crop. The audio-cassette accompanies this book and expands on some of the more complex topics. Together these form the basis of what we hope will eventually be a series dealing with all the major British crops.

The crop-specific *Handbooks* are in two sections, the first being a seasonal guide to the crop, its husbandry, pests and diseases, the second being a reference section providing detailed information about each relevant pest and disease. The video-cassette was filmed on a seasonal basis to link up with this handbook, and illustrates how to inspect the crop for different pests and diseases, and how to assess levels of infestation. At six points in the film, the student is asked to stop the tape and do an exercise based on it, using the two textbooks.

Field identification cards and recording charts are both intended for use in the field, to help monitor and record information on pests and diseases found, and on control actions taken.

We expect that the teaching packs will be used in a variety of different ways. Since they do not assume any prior knowledge of the subject, they are suitable for use directly by farmers working in their own homes. However, more often farmers will be introduced to them through study sessions set up by the Agricultural Development and Advisory Service (ADAS) or other organizations. These training sessions would be based largely on the audio and video-cassettes and farmers would then study the other components of the packs in more detail at home, keeping them for reference thereafter. The packs are also being used as supporting materials for conventional courses in agricultural training colleges, technical colleges and universities. Their advantage here is that they give much more information on a specific crop than can be included in most courses and can be used either as supporting materials for lectures or by students working in their spare time.

The cost of producing the first 1,000 packs on *Pest and Disease Management in Oilseed Rape* is outlined in Table 3.1.

TABLE 3.1

Cost of producing 1,000 teaching packs on pest and disease management on oilseed rape†

		Cost ($£$)
Course material development:	Text	10,500
	Video	10,000
Stock costs (including bought-in items) for 3 years		28,500
Promotion		5,000
Direct costs (mailing, handling, etc.)		2,600
TOTAL		56,600

† Staff costs and university overheads not included.

The Open University takes the view that good design and high quality materials are necessary for effective distance teaching. However, the same materials could have been produced more cheaply if necessary. In the British agricultural context, the cost of one pack was less than the cost of one pesticide spray round on an average farm.

CROP PROTECTION TRAINING NOW AND IN THE FUTURE

Distance teaching will become an increasingly important component in the armory of courses (FAO, 1984) for training and updating in crop protection in developed and developing countries. A common pattern of dissemination is likely to be the use of teaching packs to train advisers, who in turn use the packs to train farmers, specific components of the packs being left with the farmers for more detailed study, later reference or use in the field. An impediment to their development at the moment is the lack of appreciation in agricultural communities of the need for such training. Where this can be overcome, distance teaching provides the opportunity to reach large numbers of people very rapidly and effectively without taking them away from their workplace.

ACKNOWLEDGMENTS

The Open University is grateful to the Perry Foundation for financial support of the teaching pack program, and to the Agricultural Development and Advisory Service (ADAS) for practical support through discussions and the release of copyright materials.

REFERENCES

Corbet P.S. (1981) Non-entomological impediments to the adoption of Integrated Pest Management. *Protection Ecology* 3: 183-202.

FAO (1984) *Directory of Short Courses on Plant Protection.* Rome: FAO.

Lane, A.B. (1984a). *Understanding Crop Protection.* Milton Keynes: Open University Press, 42 pp.

Lane, A.B. (1984b), *The Oilseed Rape Handbook.* Milton Keynes: Open University Press, 70pp.

Lane, A.B. (1985). *The Winter Cereals Handbook.* Milton Keynes: Open University Press, 85 pp.

Merrill, W. (1978). Innovative teaching of plant pathology. *Annual Review of Phytopathology* 16: 239-261.

Morris, R.M. (1979). The Open University and agricultural education. *Agricultural Progress* 54: 45-56.

Morris, R.M. (1985) Agricultural Continuing Education and The Open University. *Agricultural Progress* 60: 62-69.

Tait, J.(1981). The flow of pesticides: industrial and farming perspectives. In *Progress in Resource Management and Environmental Planning, Vol.2.* eds. T. O'Riordan and R.K. Turner, Chichester: John Wiley and Sons, pp 219-250.

Tait, E.J. (in press). Rationality in pesticide use and the role of forecasting. In *Rational Pesticide Use,* eds. K.J. Brent and R. Atkin, Cambridge University Press.

Turpin, F.T. and Brown, N.H. (1979) Pest management awareness: a cartoon approach. *Bulletin of the Entomological Society of America* 25: 258-260.

4

Problems Concerning Pesticide Use In Highland Agriculture, Northern Thailand

Robert   Black,   Nuchnart   Jonglaekha,   and   Vijit Thanormthin

HIGHLAND AGRICULTURE IN THAILAND

The highlands of Northern Thailand (elevation over 800 m) cover 70% of the land area of that region. They are populated mainly by the 'hill tribes' who have had a low state of agricultural development and were unable to produce sufficient rice and other food crops to meet their needs. Traditionally, opium poppy (*Papaver somniferum*) has been the main crop. For a number of reasons, including the requirements of national security and opium suppression or replacement, there have been intensive programs of agricultural and social development in the highlands for over fifteen years. Many new crops are being introduced or are already grown by the hill tribes, including cash crops (such as coffee and temperate fruit) and improved varieties of subsistence crops. The tendency for these new crops to suffer severely from pests and diseases is a major obstacle to economic yields and high quality and hence to successful adoption by the hill tribes. The prevalence of pest problems and other contributing factors have led to widespread pesticide use in highland agriculture, and several undesirable chemicals have been introduced.

There is concern about indiscriminate pesticide use in general, and some particular conditions in the highlands increase the need for rational use of pesticides:

i)   The low level of education and agricultural development of the hill tribes means that pesticides present a special hazard to farmers, their families and livestock. Inappropriate choice of pesticide, and incorrect rates, methods and timing of application reduce their effectiveness and damage crops. It is therefore doubtful whether the hill tribes can use pesticides efficiently and safely. For example, leaf rust (*Hemileia vastatrix*) of Arabica coffee is a very serious disease in the highlands. Chemical control by regular, repeated applications of copper oxychloride is environmentally acceptable, but hill tribe farmers have generally failed to implement fully the recommended spraying program, usually leading to more severe disease levels than would occur without spraying. Such problems are not entirely due to lack of interest or ignorance as water may sometimes not be available for high volume spraying;

ii) The highlands of Northern Thailand are major watersheds, supplying populations of the north and central regions, and there is a risk of pollution of water supplies and fisheries by pesticides;

iii) There is concern about the possibility of pesticide residues in highland produce, affecting marketability within the country and abroad, although as yet no analytical data are available.

The work reported here was done under the Highland Plant Protection Program (HPPP, Chiang Mai University and Department of Agriculture). This work is mainly associated with the Royal Project, a non-government organization under Royal patronage, but also involves other agencies operating in the highlands. In practice, the Royal Project has close links with many government departments, since university staff and other government officials have carried out most of the research and development of highland crops on behalf of the Royal Project, which then takes over in the extension phase. The problems of pesticide use discussed here were encountered during the provision of a highland plant protection service.

UNDERLYING FACTORS AFFECTING PESTICIDE USE

Problems are created by the free availability of many pesticides in Thailand. Over 300 pesticides are marketed, including many which are banned or subject to restrictions in other countries (Rushtapakornchai and Vattanatungum, 1981; Staring, 1984). Agrochemical companies and retailers actively promote pesticide use, and the market is confused by the proliferation of brands for each active ingredient and the sale of unregistered products with poor labelling.

Highland agriculture operates from a broad base, involving at least five different government ministries or central government departments and many more line agencies, in addition to private foundations and commercial organizations (Table 4.1). All programs involve provision of pest management information to some extent, and also the distribution of pesticides. However, there are considerable differences in policy towards pesticide use, usually reflecting differences in emphasis on subsistence or cash crops. This compounds difficulties created by the free market for pesticides. Direct sales by agrochemical companies or retailers to farmers is also becoming an increasingly important source of pesticides in the highlands.

The nature of the crops being promoted in the highlands also affects pesticide use. These crops include rice, maize, wheat, potatoes, grain legumes, off-season and exotic vegetables, shiitake mushroom, temperate and sub-tropical fruit and nut trees, soft fruit, coffee, tea, chrysanthemum tea, cut flowers, dried flowers, pot and bedding plants and seed crops (including true seeds, bulbs and tubers).

TABLE 4.1

Organizations involved in highland agricultural development

Royal Thai Government	
Central government authority:	*Operating agency or departments*
Office of Narcotics Control Board	Highland Agricultural Marketing & Production Project Thai-German Highland Development Program Highland Coffee Research & Development Center
Ministry of Agriculture and Cooperatives	Department of Agriculture Department of Agricultural Extension Department of Land Development Royal Forestry Department Northern Region Agricultural Development Center
Bureau of University Affairs	Kasetsart University Chiang Mai University Maejo Institute of Agricultural Technology
Ministry of Interior	Public Welfare Department Police Department (Border Patrol Police Division)
Ministry of Defense	Royal Thai Army 3rd Region
Non-governmental agencies:	
Royal Project (works with many government departments) Private foundations and charities Commercial organizations (direct contracts with farmers)	

† Several projects may be operated by any one agency, and many involve financial and technical cooperation from agencies of the United Nations and the Governments of Australia, Canada, Denmark, Federal Republic of Germany, Japan, Netherlands, New Zealand, UK and USA.

Most of the cash crops are exotic, requiring practices and techniques which are sophisticated by hill tribe standards, although normal by lowland Thai standards. For example, newly introduced temperate and sub-tropical fruit (including apple, pear, peach and Japanese persimmon) must be bagged to protect them against fruit flies. This is not necessary with the native peach and apricot as they are picked green, for pickling or salting.

Pests and diseases can be serious because crop varieties are not well adapted to the area. In the research and development phase, reliance has been placed on pesticide use, without considering the implications for hill tribes. When pesticide use has been carried over to the extension phase, extension officers have not been able to evaluate the soundness of recommendations received from researchers. Coffee is attacked by stem boring and bark eating beetles (Coleoptera) and current preventative measures include the use of dieldrin. Apart from the dangers of using this chemical, it is now clear that the Thai pest species are not the same as those found in Kenya, where the recommendations originated, and further biological studies are required. Aldicarb (formulated as Temik 10G) was provided at some highland agricultural stations for use in seed potato crop trials, to provide systemic protection against aphids during early growth. Formulated as granules and incorporated into the seedbed under the supervision of qualified staff this chemical does not normally present a hazard. However, because of its very high toxicity (LD_{50} 0.9 mg/kg body weight) there is a risk of accidents if it is used by the hill tribes who, for example, often walk over fields in bare feet, and we have recommended that it should not be used under these circumstances.

PESTICIDES USED IN THE HIGHLANDS

Between June and September, 1984, the HPPP carried out a survey of pesticides stocked at 17 highland agricultural stations (mainly extension centers with some research activity) and these are listed in Tables 4.2-4.4, in order of toxicity of active ingredients. The insecticides and related products included many regarded as extremely or highly dangerous and subject to restrictions elsewhere (Table 4.2). The fungicides (Table 4.3) are less dangerous, but a number of systemic preparations were being used routinely instead of protectants, presenting a risk of resistance in target organisms. Some fungicides and insecticides were available in three or more formulations and farmers are confused by the multiplicity of brand names. Herbicides (Table 4.4) have caused most problems in use because of persistent residues affecting successive crops, spray drift and poisoning of workers (especially with paraquat which is regarded as highly hazardous by dermal exposure).

TABLE 4.2
Pesticide survey in highland agriculture: insecticides, acaricides and nematicides

	Hazard level†	Number of brands	Number of stations (Total 17)
Aldicarb	ext. high	1	1
Disulfoton	..	1	1
Mevinphos	..	1	4
Carbofuran	high	1	1
Deltamethrin	..	1	1
Dieldrin	..	2	3
Methiocarb	..	1	2
Methomyl	..	3	11
Methyl parathion	..	1	2
Monocrotophos	..	3	17
Omethoate	..	1	2
Aldrin	moderate	2	4
Binapacryl‡	..	1	1
Carbofuran	..	1	2
Dimethoate	..	2	2
Endosulfan	..	1	1
Fenvalerate	..	1	1
Heptachlor	..	1	5
Phenthoate	..	1	3
Pirimicarb	..	1	1
Quinalphos	..	1	1
Carbaryl	slight	1	11
Dicofol	..	3	5
Malathion	..	2	4
Permethrin	..	1	3
Propargite	..	1	1
White oil	—	1	1
Bacillus thuringiensis	—	3	3

†   Hazards according to standard criteria for mammalian toxicity (Oudejans, 1982).

‡   Used here as an acaricide.

TABLE 4.3
Pesticide survey in highland agriculture: fungicides, antibiotics and fumigants

Common name	Hazard level†	Number of brands	Number of stations
Methyl bromide	high	1	1
Pyrazophos	moderate	1	3
Benomyl	slight	1	7
Benalaxyl + Mancozeb	..	1	1
Bordeaux mixture + Maneb + Zineb	..	1	3
Captafol	..	1	8
Captan	..	4	6
Carbendazim	..	2	2
Carboxin	..	1	2
Chlorothalonil	..	1	5
Copper oxychloride	..	2	11
Copper oxychloride + Mancozeb	..	2	3
Copper sulfate	..	1	1
Dichloran	..	1	2
Dinocap	..	1	1
Etridiazole (Ethazole)	..	1	1
Mancozeb	..	7	12
Maneb	..	1	3
Maneb + Zineb	..	1	3
Metalaxyl	..	2	3
Oxycarboxin	..	1	5
PCNB (Quintozene)	..	2	8
Propineb	..	1	1
Streptomycin	..	1	1
Thiabendazole	..	1	1
Triadimefon	..	1	1
Triforine	..	1	1
Sulfur	..	1	2
Zineb	..	2	8

† See Table 4.2.

Note: Although ethylene-bis-dithiocarbamates (mancozeb, maneb, zineb) are only slightly toxic, there is a potential hazard because they break down to carcinogenic ethylenethiourea.

TABLE 4.4
Pesticide survey in highland agriculture: herbicides

Common name	Hazard level †	Number of brands	Number of stations
2,4-D	Moderate	2	2
2,4-D + Oxadiazon	..	1	1
MCPA	..	1	1
Paraquat	..	2	6
Alachlor	Slight	3	6
Ametryn	..	1	1
Asulam	..	1	1
Atrazine	..	2	3
Bifenox	..	1	1
Butachlor	..	1	2
Dalapon	..	2	3
Diphenamid	..	1	1
Diuron	..	1	1
Glyphosate	..	1	3
Linuron	..	2	2
Metazachlor	..	1	2
Metribuzin	..	1	1
Napropamide	..	1	1
Oxyfluoren	..	1	2
Simazine	..	1	2

† See Table 4.2

ROYAL PROJECT PESTICIDE STOCK SCHEME

The HPPP provides a routine advisory service (diagnosis and recommendations) as well as detailed supportive investigations into specific pest and disease problems, operating within the framework of IPM. In the absence of an effective central pesticide regulatory system, we have also taken specific action to control pesticide use in the Royal Project. This takes the form of a Stock Scheme whereby: orders for pesticides by extension workers using the central budget must be checked and approved by the Pesticide Coordinator, and changes made if necessary; and pesticides are drawn where possible from a stock at the Headquarters which contains sufficient products for most requirements (Table 4.5).

TABLE 4.5
Royal Project pesticide stock scheme: Products recommended by Highland Plant
Protection Program, December 1984

Common name	Trade name	Formulation †
Organochlorine insecticide		
Dicofol	Kelthane	18.5 EC
Organophosphate insecticides		
EPN	Kumiphos	45 EC
Malathion	Malarfez	83 EC
Monocrotophos	Nuvacron, Nuvaren	56 SC
Omethoate	Folimat	80 SC
Carbamate insecticides		
Carbaryl	Sevin	85 WP
Carbofuran	Furadan	3 G
Carbosulfan	Posse	20 EC
Methomyl	Lannate	90 SP
Pyrethroid insecticide		
Fenvalerate	Sumicidin	20 EC
Biological insecticide		
Bacillus thuringiensis	Bactospeine 85	8 mill iu/ml
	Argona	16000 iu/mg
Protectant fungicides		
Bordeaux mixture + Maneb + Zineb	Comac	92.5 WP
Captafol	Difolatan	80 WP
Captan	Captan	50 WP
Chlorothalonil	Daconil	75 WP
Copper oxychloride	Copper 85	85 WP
Mancozeb	Dithane M-45 Mancozeb	80 WP
PCNB	Brassicol	75 WP
PCNB + Ethazole	Terrachlor Super-X	29 EC
Zineb	Lonacol	72 WP
Systemic fungicides		
Benomyl	Benlate	50 WP
Carbendazim	Derosal	60 WP
Metalaxyl	Apron	35 SD
Oxycarboxin	Plantvax	20 EC
Pyrazophos	Afugan	30 EC
Antibiotics		
Kasugamycin	Kasumin	72 WP
Phenazine-5-Oxide	Phenazin	10 WP
Fumigants under test		
i) Soil treatment		
Metham-sodium	Fumathane	66.4 SC
ii) Post-harvest		
Magnesium phosphide	Magtoxin	

† EC = emulsifiable concentrate; WP = wettable powder; G = granules; SC =
suspension concentrate; SD = seed dressing; SP = soluble powder.

This scheme has been successful so far. It has restricted the total range of products used and prevented the purchase of undesirable organochlorines and very toxic organophosphates. (Extension officers now rarely request organochlorines). The use of systemics has also been controlled. In the case of fungicides, protectants can be substituted except where it is necessary to eradicate disease already present. With insecticides, alternate spraying of systemic and contact types, or two different systemics, is recommended.

However, the scheme has serious limitations:

i)   It only applies to extension work in the Royal Project. The research and development activities of the crop replacement program operate independently of the central plant protection service and the HPPP has no brief to provide such a scheme for other organizations;

ii) Hill tribes purchase pesticides directly from retailers and this is being encouraged by active promotion by agro-chemical companies and dealers;

iii) Several private companies now work with the hill tribes, offering contracts for cash crops, with pesticides included in the package of inputs offered;

iv) The scheme does not yet cover herbicides which, as explained, present serious problems, but it is hoped that a weed control specialist will soon join the team.

CONCLUSIONS

For a number of reasons, the emphasis in highland agriculture in Thailand has been on pest control by pesticides, and many are used without regard to their appropriateness for hill tribes. The HPPP has taken some action to control the situation, but many problems have arisen from factors beyond their control. Action needs to be taken at higher levels, to encourage only appropriate crops and cultivation practices and to regulate pesticide availability.

ACKNOWLEDGMENTS

The Highland Plant Protection Program is funded by the UDSA Agricultural Research Service as part of a program to encourage opium replacement. The senior author was employed by the UK Overseas Development Administration, which supports plant pathology at Chiang Mai University. We wish to thank our colleagues in HPPP (Dr. Danai Boonyakiat, Ms. Vorapun Charumas, Ms. Nitaya Boonmee and Ms. Sirinthip Wongdao) who assisted with the pesticide survey.

REFERENCES

Oudejans, J.M. (1982). *Agro-pesticides: Their Management and Application.* Bangkok: Economic and Social Commission for Asia and the Pacific (ESCAP), United Nations.

Rushtapakornchai, W. and Vattanatungum, V. (1981). *Thailand Pesticides Handbook.* Zoology Division, Department of Agriculture, Ministry of Agriculture and Cooperatives.

Staring, W.D.E. (1984). *Pesticides: Data Collection Systems, and Supply, Distribution and Use.* Bangkok: Economic and Social Commission for Asia and the Pacific (ESCAP), United Nations.

5

Insecticide Production, Distribution And Use: Analysing National And International Statistics

E.J. Tait and A.B. Lane

INTRODUCTION

The units used to record pesticide production, distribution and use have been expressed in many different ways — in monetary terms (usually US$), total area treated, the rate of use per hectare, the proportion of total crop area treated, the total amount of active ingredient (usually tons). This can make life difficult for the analyst who wishes to compare data from more than one source, but it also has more important and less widely appreciated consequences.

The measurement of these statistics is of little theoretical interest — it is usually done with a purpose in mind. The unit of measurement chosen is affected by that purpose and by the value system of the analyst and unless these are clearly appreciated by others with different purposes or value systems, the significance of certain trends may be missed or misunderstood.

This paper discusses the effects of using different units of measurement on the interpretation of statistics on insecticides, in relation to the purpose and value system of the analyst. Insecticides were chosen because their use creates more actual and potential problems than herbicides or fungicides, and also because there are fewer chemical groups, with more clearly differentiated characteristics, than for other types of pesticide.

UNITS USED IN PRESENTATION OF DATA

Data Expressed in Financial Terms

A financial unit, generally US$, is used by the agrochemical industry and government agencies to describe production and trading statistics, facilitating comparisons among different sets of statistics. However, in time-series data, confusion can be caused by the frequent failure to state whether figures are inflation-adjusted. Table 5.1 illustrates the absence of any real growth in insecticide sales in the UK, after the figures have been corrected for inflation.

TABLE 5.1

Sales of insecticides to merchants in the UK (£ M)

Year	Value in Current Year	1983 Value
1974	9.4	30.0
1975	10.4	26.7
1976	16.0	35.2
1977	21.8	41.4
1978	21.0	36.8
1979	23.0	35.6
1980	22.0	28.8
1981	21.5	25.2
1982	23.0	24.6
1983	30.8	30.8

Source: British Agrochemicals Association Annual Reports.

This absence of growth is a worldwide phenomenon in developed country markets for pesticides, and the changing patterns of pesticide innovation add to the difficulties of interpreting time-series data, even if they are inflation-adjusted. As the number of pesticides already on the market has increased, it has become more difficult for companies to find new products which are a significant improvement on those already available and the number of new introductions has declined (Lewis, 1976). As older pesticides outlive their patent protection they have become cheaper. Increasing competition among companies in a low-growth market, has also led to a steady fall in the inflation-adjusted price of insecticides to the consumer. This is exemplified by the decline in synthetic pyrethroid insecticides to the status of commodities within a few years of their introduction, and well before the expiry of their patents. Research workers outside the chemical industry may fail to realise that the inflation-adjusted figures in Table 5.1 conceal a considerable increase in the physical amount of pesticide applied to the land. If the purpose of an analysis is to detect trends that could have unfavorable effects on the environment or human health, this could lead to a significant underestimate, particularly since many of the older and cheaper insecticides are either damaging to the environment, e.g. organochlorines, or potentially toxic to spray operators, e.g. organophosphates.

The relative decline in the price of insecticides has other implications for those with an interest in seeing a more rational approach to their use. The cheaper they are, the more difficult it is to persuade farmers and growers to refrain from using them on an insurance basis.

TABLE 5.2
Worldwide use of insecticides on crops in 1980

Rank	Total sales ($M)		Total sales ($/ha)		Total sales ($/ton)	
1	Fruit and vegetables	1088	Tea	28	Tea	28
2	Cotton	918	Cotton	27	Cocoa	24
3	Rice	563	Tobacco	19	Cotton	22
4	Corn	418	Cocoa	9	Tobacco	15
5	Soybeans	129	Coffee	4	Coffee	9
6	Other field crops	98	Rice	4	Rubber	6
7	Tobacco	83	Groundnuts	3	Groundnuts	4
8	Wheat	76	Corn	3	Soybeans	2
9	Sorghum	68	Fruit and vegetables	3	Rice	1
10	Tea	53	Soybeans	2	Sorghum	1

Source: Farm Chemicals, Sept. 1981 and FAO Production Yearbook.

Financial data can be combined with crop area or tonnage of crop produced to alter the information content of a unit. Table 5.2 illustrates the effect on the rank ordering of crops, of describing insecticide use in terms of total sales in $US, sales/hectare grown and sales/ton of produce. From the point of view of someone interested in toxic or environmental side effects of insecticides, tea is therefore a relatively unimportant crop on a world basis, but in those areas where the crop is widely grown it could be a very important contributor to pesticide problems. The high position of cotton on all three measures in Table 5.2 accounts for its controversial nature — it is the most important crop market for insecticides and also the most important single cause of environmental problems and toxic side-effects on agricultural workers.

Data Expressed as Tons Active Ingredient

The wide variation in potency of different insecticides can make it difficult to interpret data expressed as tons active ingredient unless the nature of the pesticides involved is also known. Organophosphate insecticides are on average twice as potent, and carbamates and synthetic pyrethroids ten to a hundred times as potent, as organochlorines. Within each class there is also a wide range of potency among individual chemicals. Braunholtz (1981) has shown how there has been a steady reduction in the rate of application of insecticide required to give adequate control of cotton pests in the USA.

This reduction in the quantity of active ingredient required per hectare has been widely referred to as an environmental benefit, but this need not be the case, unless the selectivity of the chemicals in question is also improved. None of the modern insecticides suffers from the problem created by the persistence in food chains of the organochlorines. However, most are still relatively broad spectrum compounds and their repeated application to large areas of land could have significant long term effects on wildlife. The organophosphate and carbamate insecticides are much more acutely toxic to spray operators than were the organochlorines and the increased scale of their use had lead to many more accidents and deaths, particularly in developing countries. The synthetic pyrethroid insecticides were widely promoted as being safe to the environment and to people, an image which was fostered by their origin in the natural product, pyrethrum. However, the changes to the parent molecule which gave the synthetic products their greater stability and effectiveness have also increased their mammalian toxicity to levels comparable with many organophosphorus and carbamate insecticides. They are also toxic in an aquatic environment, although not so seriously as was originally feared (Stephenson et al., 1984).

TABLE 5.3
Insecticide use on crops in the USA, 1966 to 1976

Insecticide	Quantity (M lbs.) 1966	1976	Difference
Toxaphene	30.9	31.6	+0.7
DDT	26.3	—	-26.3
Aldrin	14.8	0.9	-13.9
Carbaryl	11.8	16.5	+4.7
Parathion	8.4	10.5	+2.1
Methyl parathion	8.0	22.9	+14.9
Carbofuran	—	11.6	+11.6
Phorate	—	6.4	+6.4
Disulfoton	1.9	6.2	+4.3
Others	35.9	55.4	+19.5
Total	138.0	162.0	+24.0

Source: Eichers (1981).

Time-series data which describe pesticide production and marketing in terms of tons or pounds of active ingredient should always be interpreted with these underlying trends in mind. For example, a cursory inspection of Table 5.3 could lead one to conclude that there had been a straightforward substitution of 40.2 million pounds of organochlorine insecticides by approximately 44 million pounds of organophosphates and carbamates. However, given that the recommended application rate of DDT varies from

1 to 3 kg/ha, depending on the crop, while the recommended rates for organophosphates and carbamates vary from 0.2 to 1 kg/hg, these figures represent a very considerable increase in the area treated with insecticide over the ten year period, much more than is implied by the small increase in the total quantity of active ingredient involved. As Eichers (1981) shows, there was a 76% increase in the farm acreage treated with insecticides over the same period.

Aggregate data on insecticide production and use, expressed in terms of physical quantities, will become more meaningless as the proportion of synthetic pyrethroids in use increases and as their potency increases. Insecticide consumption is expected to increase from 1,590 thousand tons in 1980 to 2,235 thousand tons in 1995, a projected annual growth rate of 2.3% compared to an historic annual growth rate of 3%, (Anon, 1983). The synthetic pyrethroids began to take up a significant portion of the market in 1976: by 1980 they constituted 9% of foliar insecticide sales and by 1985 they were expected to have 20-25% of this market (Cox, 1981). The application rates of third generation pyrethroids are approximately 0.11 kg/ha, and for the new fourth generation pyrethroids from 0.01 to 0.06 kg/ha (Ware, 1983). This means that one ton of the third generation synthetic pyrethroid insecticide, permethrin, could give a marketable effect equivalent to three to five tons of carbamate or organophosphate and ten to 30 tons of DDT. One ton of a fourth generation pyrethroid could replace from eight to 30 tons of organophosphate or carbamate and 100 to 300 tons of DDT. The true meaning of the above growth projection will therefore depend on the relative proportions of different types of pesticide involved.

Further complicating factors are introduced when pesticide supply is measured in terms of physical quantities of *formulated* pesticide, given the very large number of possible formulations, liquid or solid, more or less concentrated. Taken alone they give little indication of the true quantities of chemicals involved or of the nature of any likely problems resulting from their use.

Data Expressed as Area Treated with Pesticide

The number of hectares of crop receiving a pesticide treatment gives little useful information about pesticide usage. The significance of such data only becomes apparent when they can be viewed in relation to the total cultivated crop area, as shown in Table 5.4. The third column in this table refers to the number of acres on which insecticides were used at least once. This does not take account of the fact that farmers who use insecticides frequently apply them more than once to the same crop. Depending on the extent to which multiple insecticide applications occur, 'acres treated' can seriously underestimate the extent of insecticide use.

This problem can be partially overcome by expressing pesticide usage as 'spray hectares', a cumulative measure of the number of hectares treated. If a large proportion of the crop is treated more than once, it is possible for the 'spray hectares' to be greater than the total area of crop grown. The

TABLE 5.4
Insecticide use on major field crops in the USA in 1976

Crop	Acres grown (M)	Acres treated (M)	Acres Treated (%)
Corn	84.1	32.0	38
Cotton	11.7	7.0	60
Wheat	80.2	11.2	14
Sorghum	18.6	5.0	27
Rice	2.5	0.3	12
Other grains	29.8	1.5	5
Soybeans	50.3	3.5	7
Tobacco	1.0	0.8	30
Peanuts	1.5	0.8	53
Alfalfa	26.5	3.5	13

Source: Eichers (1981).

unit, hectares treated is ambiguous, sometimes being used in sense referred to in Table 5.4, and sometimes as synonymous with 'spray hectares', but it is of little use in the absence of information on the potential area for pesticide treatment. As shown in Table 5.5 the number of hectares of oilseed rape treated with insecticide in the UK has risen from 15,000 to 177,000 in only six years. However, in the same period, the area of crop grown has also risen from 55000 to 222000 hectares so, although there has been a 12-fold increase in the spray area itself, the increase in proportion to the total crop areas has only been approximately three-fold, from 0.27 to 0.79. Table 5.5 also shows that, although the increase in total acreage of cereals grown has been relatively modest (19%) between 1974 and 1983, there was a 35-fold increase in the area treated with insecticide to almost one million hectares. The area of sugar beet and potatoes grown has been fairly stable over this period and the figures for 'spray hectares' show considerable fluctuations in insecticide use, with most of the crop being treated at least once in most years.

'Hectares treated' and 'spray hectares', unlike data expressed in financial terms or in terms of physical quantities, do not take account of the variation in quantity of pesticide applied which depends on factors such as leaf area of the crop, row spacing, soil type, method or volume of application. For example, the recommended rate of application for demeton-S-methyl on potatoes and sugar beet is 244 g/ha, while on wheat it is 125 g/ha. A 'spray hectare' of insecticide on potatoes or sugar beet could therefore involve almost twice as much insecticide as on cereals.

TABLE 5.5
Insecticide usage on major crops in the UK ('000 ha.)

	1974	1975	1976	1977	1978	1979	1980	1981	1982	1983
Cereals										
Area grown	3390	3414	3154	3209	3757	3792	3873	3918	3908	4036
Spray ha.	28	59	772	564	100	311	440	536	406	977
Potatoes										
Area grown	188	172	149	177	199	189	190	162	174	181
Spray ha.	177	182	200	277	167	191	139	80	71	114
Sugar beet										
Area grown	195	—	—	202	210	213	212	206	210	199
Spray ha·	311	—	—	182	196	297	113	236	90	236
Oilseed rape										
Area grown	24	—	—	55	—	74	92	124	170	222
Spray ha.	17	—	—	15	—	26	30	43	89	177

Source: British Agrochemicals Association Annual Reports.

Standardized Pesticide Usage

    'Standardized pesticide usage' was developed as a method for making comparisons at the micro level, for example, to explore detailed differences in the pesticide usage profiles of farmers (Tait, 1977; Tait, 1983) or to study social, psychological and economic factors influencing pesticide usage (Tait, 1978; Tait, 1983). The process of standardization removes fluctuations in pesticide usage data which are attributable to biological, chemical or physical factors like the effects of weather on pest incidence at different times in the growing season, the variation in potency of different pesticides, and the variation in application rates for different growing conditions and different crops. The residual variation in pesticide usage, within and between farms, provides a key to the investigation of behavioral trends and anomalies in the farming population, their causes and possible means of encouraging changes in behavior, where this is considered desirable.

    The starting point for such an analysis is the collection of detailed information on pesticide usage from a sample of farmers who grow the crops in question. The area being studied is divided into regions, within which there is assumed to be no consistent variation in pest problems, although random fluctuations in pest incidence from one farm to another are always bound to occur. Pesticide usage data are collected from a random sample of farmers in each region and are coded in 'units', a unit being defined as the quantity of pesticide used when the farmer applies it to his total crop acreage at the rate recommended by the manufacturer. If the farmer applies pesticide to less than his total crop acreage or uses more or less of the chemical than is recommended by the manufacturer, or applies more than one pesticide in a tank mix, the units are weighted accordingly.

The mean number of pesticide units used on each crop subdivision is then calculated and each farmer's pesticide use is expressed as a standardized deviation from this mean. As shown in Figure 5.1, this results in a 'pesticide usage profile' where the mean pesticide usage has the value zero, and departures from the mean are measured in standard deviations. If all the farmers in the sample had applied insecticides for aphids and caterpillars on vegetable brassicas according to strict scientific rationality, deviations from the mean insecticide usage would have been randomly distributed. The consistency of pesticide usage within farms, and the significant differences between farms, indicate deviations from such rational use which can be attributed to some extent to social and psychological influences. 'Standardized pesticide usage' thus constitutes a behavioral index against which the many possible factors influencing farmers' behavior can be correlated. (Tait, 1978; 1983).

Pesticide usage can be standardized at various levels of aggregation, e.g. individual insecticides, insecticides in general, insecticides used for particular pests, or all pesticides used on a crop, depending on the purposes of the investigator. If the focus of interest was on control of a particular pest or pests, it would be appropriate to look at insecticide use for those pests only; if the focus of interest was on a particular chemical or group of chemicals, attention could be restricted to them; if the aim was to study how farmers make selections among insecticides, then all insecticides available to them should be included in the analysis.

CONCLUSIONS

The methods of measuring pesticide production and use described here have been widely used at the regional, national and company levels. They have also been employed at the farm level where pesticide use is described in terms of money spent per hectare for financial analyses, or in terms of physical quantities of active ingredient, or area treated for a chemical or biological analysis. 'Standardized pesticide usage' is restricted to farm-level analysis where the focus of interest is on variables of a social scientific nature.

This paper has described how the choice of a measure reflects the aims and interests of the analyst and has also indicated some of the biases that are inevitably introduced in each case. There is no such thing as an ideal measure and, if time and available data permit, it is preferable to present statistics in a variety of different forms.

The potential pitfalls and constraints associated with each measure can be summarized as follows:

- time series data expressed in financial terms should always state whether they have been adjusted for inflation;

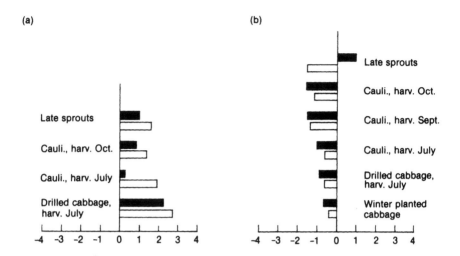

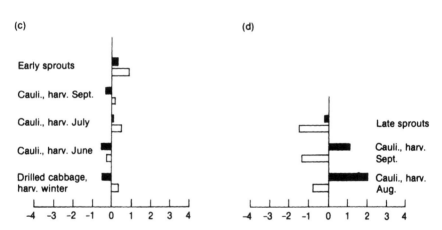

Figure 5.1   Farmer pesticide usage profiles for Brassica crops in Lincolnshire, UK.
■ Insecticide usage for aphids;   □ insecticide usage for caterpillars

- time series data expressed in financial terms will tend to underestimate the increase in the physical amounts of insecticide produced or used, due to the fall in insecticide prices in recent years;

- data expressed in terms of quantity of active ingredient should not be aggregated across chemical groups with a wide variation in potency or other characteristics;

- in interpreting insecticide statistics where data have been aggregated across chemical groups, increases in the amount of toxicant added to the environment will be underestimated due to the introduction of more potent chemicals in recent years;

- information on acreage treated with insecticide should always be accompanied by the relevant total crop acreage;

- data expressed as hectares treated or 'spray hectares' do not take account of the considerable variation in the amount of insecticide used, depending on the crop in question;

- 'standardized pesticide usage' gives no indication of the acreage treated by pesticide or of the quantities of active ingredient used.

The use of insecticides attracts strong personal opinions from protagonists and antagonists, and it is not uncommon to find units being chosen for their dramatic effect rather than their contribution to knowledge. Thus, financial statistics emanating from the agrochemical industry are frequently not adjusted for inflation, thereby enhancing the apparent growth in sales. On the other side of the divide, environmental and other anti-pesticide pressure groups have a tendency to quote figures on the acreage of crops treated with pesticide or the aggregate quantity of pesticide active ingredient used, without making due allowance for the difference in characteristics of different chemical groups.

Those with an interest in dispassionate analysis can ensure that their own measurements are carried out in the most appropriate units, that the meaning of the data and the biases introduced by the measure are made clear to others, and that they are sufficiently critical of data from other sources which may have concealed biases.

REFERENCES

Anon. (1983). Mixed World Demand for Pesticides Predicted. *GIFAP Bulletin* 9(11): 8

Braunholtz, J.T. (1981). Crop protection: the role of the chemical industry in an uncertain future. *Philosophical Transactions of the Royal Society of London B* 295,: 19-34

Cox, J.R. (1981). Pyrethroids in the 1980's. In *Agrochemicals Outlook for the 1980's,* Proceedings of the First Wood Mackenzie Conference on the Agrochemical Industry, Wood Mackenzie & Co.

Eichers, T.R. (1981). Use of Pesticides by Farmers. In *CRC Handbook of Pest Management in Agriculture*, ed. D. Pimentel, Lewis 1976 Vol. 2, pp 3-25.

Lewis, C.J. (1976). The economics of pesticide research. In *Origins of Pest, Parasite, Weed and Disease Problems* pp. 237-245. Proceedings of the 18th symposium of the British Ecological Society.

Stephenson, R.R., Chou, S.Y. and Olmos-Jerez,A. (1984). Determining the toxicity and hazard to fish of a rice insecticide. *Crop Protection* 3(2): 151-165.

Tait, E.J. (1977). A method for comparing pesticide usage patterns between farmers. *Annals of Applied Biology* 86 : 229-240.

Tait, E.J. (1978). Factors affecting the usage of insecticides and fungicides on fruit and vegetable crops in Great Britain: II Farmer-specific factors. *Journal of Environmental Management* 6 : 143-151.

Tait, E.J. (1983). Pest control decision making on brassica crops. In *Advances in Applied Biology*, ed. T.H. Coaker, Vol.8, London: Academic Press, pp. 122-188.

Ware, G.W. (1983). *Pesticides: Theory and Application*. San Francisco: W.H. Freeman & Co.

6

A Case Study Of Pest Management On Cotton In Queensland

J.P. Evenson

INTRODUCTION

An informal research group in the Agriculture and Entomology Departments at the University of Queensland, interested in the problems of pest management in cotton, was formed in 1975, funded by the Australian Research Grants Committee. The group developed an alternative strategy for growing cotton, employing a more ecologically sound approach to pest control than any currently available (Blood et al., 1975). This alternative strategy was based on overcoming environmental, technological and economic barriers.

In contributing to the environmental barrier, conventional pest control occasions the use of either persistent, slowly-degradable compounds, or rapidly degradable but highly toxic compounds. The technological barrier refers to extension problems whereby any new techniques would have to integrate into existing agricultural practices and to be effective in controlling pest populations within prescribed time limits. To overcome the economic barrier any alternative strategy, must compete effectively with the benefit:cost ratio of conventional control schemes (Longworth and Rudd, 1973).

The group first investigated cotton ecosystem dynamics in the absence of pesticides. Over five consecutive seasons it proved possible to grow crops in the absence of pesticides and produce a commercial yield, but the risks of failure remained unacceptable (Bishop and Blood, 1977; 1978; Bishop et al., 1978)

During the 1975/76 season it was decided to experiment with a heuristic model of control on a large scale. Professor W.L. Sterling, Texas A & M University, who had valuable experience in developing and implementing successful pest management programs in Texas, helped to develop this model. The group's objective was to use the control dynamics already existing in the unsprayed system and at the same time decrease the risk of system failure by incorporating extrinsic biocides harnessed to an efficient decision management frame-work.

The choice of extrinsic controls was governed by the 'barriers' described above, environmental integrity, technological effectiveness and economic efficiency, and unfortunately also by what was 'available' from the potential control techniques listed by Longworth and Rudd (1973).

The *Heliothis* complex was judged to be the key insect problem and therefore supplementary ovicides and larvicides were sought which overcame the three above barriers. Chlordimeform, a formamidine compound, was chosen as the ovicide because of its low mammalian toxicity, low persistence in the environment and its reportedly minor effect on beneficial insects and spiders. It could be applied quickly and was competitive in cost-effectiveness with conventional insecticides. A *Bacillus thuringiensis* (B.t.) formulation was chosen as the larvicide because of its pest specificity, environmental safety and its synergism with chlordimeform. The only drawback was its high cost. The group would have preferred to use biological methods to induce egg mortality (such as the use of egg parasites like *Trichogramma* and *Telenomus*) but submissions to funding authorities for the establishment of parasite production facilities met with no success.

Any successful new strategy must possess long-term stability and yet retain sufficient flexibility to deal with unpredictable exigencies. In cotton pest management, long-term stability can only be guaranteed by the absence of the potential for the development of insecticide resistance. Any system relying wholly on conventional compounds runs this risk. Chlordimeform, although a relatively new compound without any reported resistance, was vulnerable to this criticism. Also, although resistance to B.t. has not yet arisen, it cannot be categorically ruled out. Although the two materials were the most suitable available, their rational use had to be considered in the context of the management framework which decided when and in what manner to employ them. Thus the management framework, the control compounds and the intrinsic beneficial insects in the system formed the integrated pest management package.

The management framework incorporated the best available information on treatment thresholds, together with the most efficient method of monitoring and sampling.

Success with sequential sampling in Texas, E. Africa and California suggested that the group should employ this approach rather than conventional methods, because of its speed of operation and its ability to classify as well as estimate populations. The fact that sequential sampling programs can be tailored for specific farm situations and implemented by relatively untrained operatives reinforced our decision (Sterling, 1976).

The group felt that a frontal assault on the current pest control methods would meet with resistance from farmers who would feel that their livelihood was being threatened. It was therefore argued that any pest management system would have to develop incrementally recognizing four phases of development:

i) the natural biological stage, representing the natural bio-control system in the absence of chemicals, which must be understood before it is possible to produce an improved management system;

ii) the commercial chemical stage representing the current state of the art;

iii) the commercial mixed stage representing a planned development of pest control based on enhancing natural biocontrol by any suitable means, including chemicals, to achieve a commercially acceptable intermediate system;

iv) the commercial biological stage representing the final objective of total biological control of a system.

By 1979 considerable progress had been made in the development of a pest management system for cotton in Queensland. Table 6.1 compares the results of four years' field trials on a 50 ha 'pest management' block with a block under normal commercial management on the same farm. (Since 1976/77 all decisions were based on sequential sampling.)

TABLE 6.1

Comparison of cotton yields and spraying regimes in the Lockyer Valley

Systems used:	Pest Management†				Commercial‡			
Harvest year:	76	77	78	79	76	77	78	79
No. of sprays	20	12	10	11	17	14	9	12
Total insecticide load (kg ha)		4.0	6.0	6.0	9.8	50.0§	32.0*	16.0*	16.9*
Total insecticide cost ($ ha5-14)	178	84	74	95	165	116	78	106
Yield (bales ha5-14)		3.5	4.9	4.0	5.6	4.5	4.7	3.4	5.0

§ Farmers' own decisions on spraying

* Sequential sampling used for decision making

Main pesticides used:

† chlordimeform, amitraz, NPV (nuclear polyhedrosis virus), *B. thuringiensis*, endosulfan.

‡ endosulfan, monocrotophos, dimethoate, D.D.T./camphechlor.

Table 6.2 compares yields obtained from pest management systems representing stages (ii) (iii) and (iv), indicating a reasonable basis for suggesting that improvement in pest management efficiency was possible. Using only some elements of pest management, yields can be maintained or increased and the total insecticide load on the environment can be reduced without increasing insecticide costs. In the commercial system, the value of decision making using sequential sampling was also demonstrated, insecticide load and cost having been reduced since 1975/76.

By 1978 many Queensland cotton farmers had adopted parts, if not all, of the pest management approach as a result of information from this project. Waite also demonstrated the potential of the system in a more difficult environment, at Emerald (pers. comm.). Farmers had adopted the sequential sampling system of decision making, which laid stress on maintenance of beneficial insects using endosulphan, the only commercially available chemical that had less serious effects on beneficial insect populations.

TABLE 6.2

Cotton yields, pest control costs and pesticide loads, Lockyer Valley, Queensland 1977/78

	Yield (kg. ha^{-1})	Cost ($ ha^{-1})	Pesticide Load (kg. a.i. ha^{-1})
Commercial stage system (100 ha)	3.4	78.0	16.0
Mixed stage system (40 ha)	4.1	74.0	6.0
Bacteria-based † experimental system (8 ha)	5.4	70.0	9.0
Virus based † experimental system (8 ha)	4.1	81.0	5.0

† Both systems need some chemical sprays, usually as low rate ovicides.

Although they had demonstrated a willingness to change, they were hampered by lack of biological control agents at a reasonable price and by application systems that were not highly efficient. Nevertheless a significant change in farmer attitudes was evident with requests for training on sequential sampling methods, discussions on damage thresholds, and for the supply of trained independent scouts.

PESTICIDE USAGE IN THE QUEENSLAND COTTON INDUSTRY IN THE 1978/79 AND 1979/80 SEASONS

During the 1978/79 and 1979/80 cotton seasons a survey was conducted to determine the factors operative in changing farmer attitudes to pest management. This survey, to be reported elsewhere, also gathered baseline data on the methods used for dealing with pest problems. Those interviewed represented 92 percent of all farmers in the Queensland Industry. The results of the survey follow.

The insecticides used are listed in Table 6.3. DDT/camphechlor was used only as a mixture. All the others were available singly but were often used in combination.

TABLE 6.3
Insecticides used on cotton in Queensland 1978/79 season

Organochlorine	Organophosphate	Pyrethroid	Other
DDT	parathion-methyl	fenvalerate	chlordimeform
endrin	profenofos	permethrin	
DDT/	omethoate	decamethrin	
camphechlor †	dimethoate		
endosulfan	monocrotophos		
	methomyl		
	demeton-S-methyl		

† Farmers indicated that they were turning away from DDT/camphechlor and in 1982 it was banned from use on cotton.

Table 6.4 lists the number of spray rounds (one spraying for a particular block) for the most widely used insecticides. The percentage of spray rounds in which combinations were used ranged from 3% for endosulfan to 18% for fenvalerate and 19% for DDT/camphechlor. Table 6.5 shows the usage over all regions of the three most important insecticides.

Application Method

Only 17 spray rounds used ultra low volume (ULV) formulations, 16 of fenvalerate and one of methomyl. Tractor mounted ground rigs were used by more than 50% of farmers early in the season until the crop became too dense or the weather prevented wheeled vehicle movement. Aircraft were the main means of applying chemical in mid to late season. Most used low volume application by fixed wing aircraft.

TABLE 6.4

Number of spray rounds for the main chemicals used and mixtures containing these chemicals

Chemical	No. of events	Chemical	No. of events
endosulfan	549	DDT/camphechlor	113
+ chlordimeform	7	+ parathion-m	17
+ methomyl	8	+ monocrotophos	5
+ monocrotophos	1	+ profenofos	1
fenvalerate	246	+ methomyl	1
+ parathion-m	26	+ chlordimeform	1
+ dimethoate	6	+ demeton-s-methyl	1
+ monocrotophos	3		
+ profenofos	2		
+ omethoate	1		

TABLE 6.5

Usage of pesticides by region (% of spray rounds in which chemicals were used either singly or in mixtures)

Region	endosulfan	fenvalerate	others
Emerald	26	61	13
Biloela	71	21	8
Theodore	90	5	5
St. George	51	11	38
Downs & Lockyer	30	21	49

† Largely DDT/camphechlor

Target Insects

Table 6.6 lists the insects targeted in specific sprays, showing *Heliothis* to be the main pest. The critical point to observe in spraying *Heliothis* was that *H. punctiger* always appeared first and was controllable by endosulfan whereas *H. armiger* which usually appeared almost half way through the season was not. The use of trained scouts who could detect the change enabled farmers to use endosulfan early in the season to avoid destroying parasites and predators. Adoption of this policy was responsible for differences in district performances in pesticide use (Table 6.5). Where analysis could be performed, pyrethroid users had more problems with aphids and thrips (Table 6.7).

TABLE 6.6

Percentage of spray rounds targeted on specific insects or groups (for first nine sprays only)

	Spray number								
	1	2	3	4	5	6	7	8	9
Heliothis only	64	86	87	86	84	71	65	49	37
Heliothis & others †	6	4		5	3	5	5	6	2
Aphids, mites, jassids, mirids, thrips	24	4	1	1	1	1	4	6	6
Tipworms	5	4	3	2	1	1	—	—	—
Pink spotted bollworm	—	—	—	1	2	2	5	4	4
No spray applied	1	2	4	7	7	20	20	39	48

† 'others' include tipworms, jassids, aphids & rough bollworm

TABLE 6.7

Percent of spray rounds using pyrethroids compared with percent of farmers in the same district spraying for thrips and aphids

Location	% Sprays containing pyrethroids	% Sprays targeted on aphids and thrips
Emerald	40.9	11.47
Downs	19.75	9.06
St. George	11.08	7.35
Biloela	12.80	1.83
Theodore	2.33	1.33

Note: Regression analysis on transformed data gave $R^2 = 0.72$

CONCLUSIONS

The marked regional differences in pesticide usage could be attributed to a difference in: a) pest spectrum; b) chemical effectiveness; c) attitudes to pest management; d) information services available to farmers. The first possibility may have some influence but the second cannot be supported (Waite pers. comm.). The latter two suggestions need to be examined more closely, but results from one district showed that farmers hiring independent scouts applied an average of 8.4 sprays while those using chemical company advice used 10.7 sprays.

The survey showed that, in 1979/80, 89 of the farmers interviewed could be classified as using or attempting to use the pest management approach, whereas 38 used broad spectrum chemicals only. There was no significant difference between the mean lint yields of the two groups, but there was a significant difference in the mean number of sprays used by farmers (8.7 for those adopting the pest management approach and 10.9 for those using chemicals only).

The average cotton area per farm was 112 ha and costs were estimated at $32.50/ha for two sprays and $12.36/ha for scouting services. A net industry saving of over $200,695 was therefore achieved in 1979/80 alone for the 89 farmers, equivalent to $2255 per farm (or $20.14/ha). The total research grant received for pest management research at the University of Queensland was $250,000 over the period 1973-1978 ($50,000/year), clearly demonstrating the cost effectiveness of the research and subsequent extension.

Since 1981, a computerized insect pest management system has been undergoing trial and extension in Queensland after previous testing and use in New South Wales (Australian Cotton Grower, January, 1981 p.46). Given the experience in Queensland of using sequential sampling (Sterling, 1976), the system cannot fail to produce reductions in total application and in farmer profit in areas where no such system has operated before.

ACKNOWLEDGMENTS

The author wishes to acknowledge the field work done by Lindy Capon without whose efforts the survey could never have been carried out. Drs B. Crouch and Shankariah Chamala participated in the project with special reference to extension and socio-metric aspects of information flow not reported here.

REFERENCES

Bishop, A.L., and Blood, P.R.B. (1977). A record of beneficial arthropods and insect diseases in southwest Queensland Cotton, *Pest Articles and News Summaries* 23(4): 384-386

Bishop, A.L. and Blood, P.R.B., (1978). Temporal distribution and abundance of the coccinellid complex as related to aphid populations in southeast Queensland. *Australian Journal of Zoology* 26: 153-158.

Bishop, A.L., Day, R.E., Blood P.R.B. and Evenson, J.P. (1978). Distribution of cotton looper (*Anomis Flava* Fabr.) larvae and larva damage on cotton and its relationship to the photosynthetic potential of cotton leaves at the attack sites. *Australian Journal of Agricultural Research* 29: 319-325.

Blood, P.R.B., Longworth, J.W. and Evenson, J.P. (1975). Management of the cotton agroecosystem in Southern Queensland: a preliminary modelling framework. *Proceedings of the Ecological Society of Australia* 9: 230-249.

Longworth, J.W. and Rudd B. (1973). Plant pesticide economics with special reference to insecticides. *Australian Journal of Agricultural Economics* 9: 210-227.

Sterling, W.L. (1976). Sequential decision plans for the management of cotton arthropods in Southeast Queensland. *Australian Journal of Ecology* 1: 265-274.

7

"Letters To The Editor" And The Perception Of Weed Control Strategies: The Use Of 2,4-D To Control *Myriophyllum spicatum* L. In The Okanagan Valley, British Columbia, Canada

Philip Dearden

INTRODUCTION

Responses to hazards are controlled more by perceptions of the hazard events than the objective reality of those events (White, 1945; Kates, 1962; 1978). The hazard under investigation here is technological, the application of the herbicide 2,4-D to control an infestation of the aquatic weed Eurasian water milfoil (*Myriophyllum spicatum*). Although 2,4-D was perceived as a hazard by some, it was not by others, particularly the agency applying the herbicide (Dearden, 1984). However, such was the strength of opposition to its use that the agency program was eventually rendered ineffective through prolonged delay brought about by program opponents. Considerable resources in the form of agency time, money and effort were therefore directed toward a program that failed to realise the desired results.

Throughout the dispute between those for and against the use of 2,4-D, considerable use was made of the media, in particular local newspapers, to voice each side's point of view. This paper reports on one segment of this interaction, the forum created by the "Letters to the Editor" section of local newspapers. The central question is whether the letters to the editor provide a valid reflection of public opinion and if so, could monitoring of this source of information have warned decision-makers of the strength of opposition to the program.

A content analysis was undertaken of all letters appearing in the "Letters to the Editor" section of local newspapers. Information was classified by newspaper, date, source, source affiliation, target, overall position for or against 2,4-D and specific attitudes expressed. The results of this analysis were compared to the results of a mailed questionnaire sent to homes randomly selected from telephone directories within three communities in the area (Dearden, 1983).

THE INFESTATION AND CONTROL PROGRAM

Eurasian water milfoil was first noticed in the Okanagan Valley, British Columbia, in 1971 by concerned citizens who found the water adjacent to their favorite beaches to be infested, interfering with swimming. The species is not native to North America and many authors have described the invasion of milfoil into previously uninfested waters (Coffey and McNabb, 1974; Smith, 1971) The weed is a rooted, perennial, aquatic, macrophyte principally occupying lakeshores. The rate of expansion through the Okanagan system after initial colonisation is illustrated in Figure 7.1.

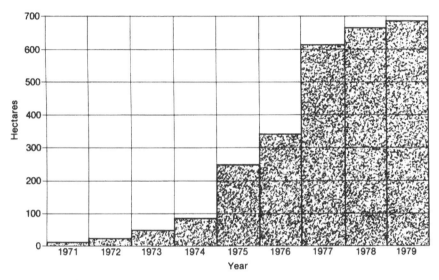

Figure 7.1   Growth of milfoil in Okanagan Valley

The rapid expansion caused milfoil to be seen as a hazard in three respects (Figure 7.2). First, it displaced native species and caused ecosystem changes affecting shore-spawning fish species and waterfowl (Water Investigations Branch, 1980). Second, water management practices were made more costly as milfoil caused flooding, impeded drainage and clogged pumps and filters. This was particularly important in the Okanagan due to the large amount of irrigated agriculture (approximately 35,000 ha in 1984). Finally the milfoil had a severe impact on water-based recreation. Eighty percent of respondents to a mail questionnaire stated that they had reduced their swimming activities as a direct consequence of the milfoil infestation, 59% had reduced boating and 41% fishing. Furthermore, 42% of lakeshore property owners estimated that their property value had declined because of milfoil.

These figures suggest a potentially significant impact upon the tourist trade. Over $200 million per annum is generated by tourism, the area's

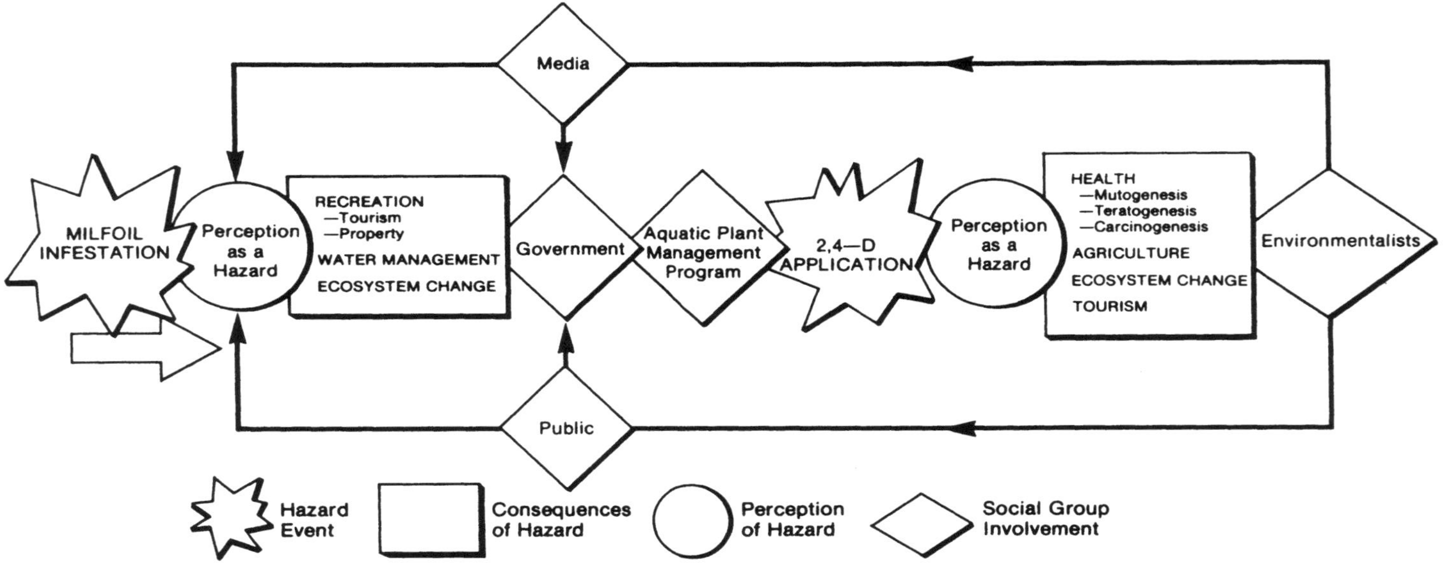

Figure 7.2   Flow diagram of the milfoil and 2,4–D hazards

largest industry,and the main attraction for tourists is water-based recreation (O'Riordan and Collins, 1974). There is evidence that the infestation adversely affected tourism in the area, although it is difficult to place a monetary value on the loss (Dearden, 1984).

As a result of this perceived threat to the local economy, the government created a special agency, the Aquatic Plant Management Program (APMP) to address the issue (Figure 7.2). After investigating many possible control procedures (Newroth, 1979) APMP suggested that 2,4-D should form part of an integrated program to overcome the problem. This suggestion and its subsequent implementation generated much controversy leading to physical obstruction, legal challenges and legislative change. The issue dominated regional media coverage for almost a decade and gained wide publicity throughout British Columbia and in many tourist areas in Canada and the western United States. The anti-2,4-D forces claimed that the herbicide constituted a hazard to ecosystems, tourism, agriculture (particularly grapes) and health (Dearden, 1984). On the other hand the pro-2,4-D forces, particularly the APMP, refused to acknowledge the existence of such concerns and continued to try to implement the control program.

Without prejudging which view of 2,4-D is 'correct', the expensive confrontation between the two sides was unproductive in terms of achieving control of the weed and was an inefficient use of public funds. APMP made no assessment of public perceptions of the milfoil or 2,4-D problems. This paper examines one aspect of public perceptions, those revealed by "Letters to the Editor" in Okanagan Valley newspapers.

LETTERS TO THE EDITOR

The number of letters over time, for and against the use of 2,4-D, published by Okanagan Valley newspapers is shown in Figure 7.3. Of the total of 133 letters, 100 were against the use of 2,4-D and 33 in favor. This contrasts with the results of a mailed questionnaire which indicated that 54% of the 403 respondents were in favor of 2,4-D, where "it is the most economical control method, if approved by the Pesticide Control Branch" (the exact wording of the question). This suggests that APMP had a majority of public support for 2,4-D application and that the letters to the editor did not provide an accurate impression of public opinion. APMP took this point of view, pointing to the low attendance at public meetings as an indication of lack of public opposition to their policies. However, several other factors need to be taken into account before endorsing such conclusions.

The comparison between the letters and the questionnaire calls into question the validity and reliability of each as a reflection of public opinion. The questionnaire results were based upon a larger data base (403 respondents out of 1,500 in the sample). Although the majority of respondents endorsed the use of 2,4-D, the herbicide was also the *least* preferred method of control overall, ranking behind harvesting, biological and dredging

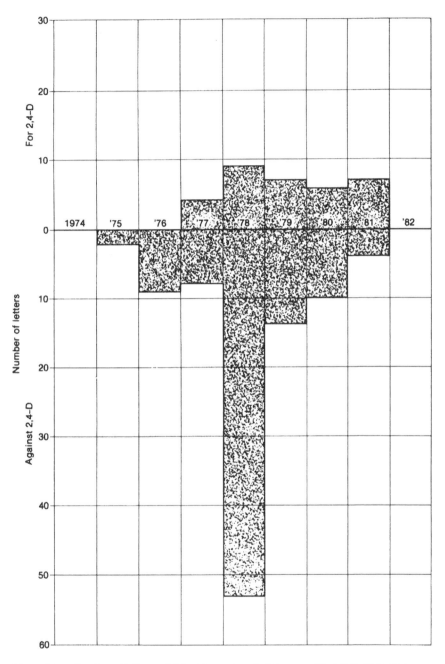

Figure 7.3   Number of letters published for and against the use of 2, 4-D in Okanagan Valley newspapers, 1974–1981

techniques respectively. Furthermore, 2,4-D was seen in a less favorable light than other commonly used chemical treatments such as agricultural pesticides (where 70% of the respondents favored their use), chlorination of water (70%) and fluoridation of water (61%). Thus although a majority of respondents approved of the use of 2,4-D, this did not indicate a solid endorsement for this course of action.

In studying the representativeness of letters to the newspapers as indicators of public opinion, several studies have found letter writers to constitute a wealthier, better read, older and more highly educated population (Haskins, 1967; Buell, 1975; Volgy et al., 1977). Political commentary has characterized letter writers in the U.S. as being largely conservative and Republican (Grey and Brown, 1970; Renfro, 1979), although some authors have suggested that letter writers are fairly representative of public opinion (Rosenau, 1974). Furthermore, some newspapers exert firmer editorial control than others on the letters that are printed so that the letters printed are not necessarily representative of those that have been written. Nonetheless, a recent paper by Hill (1981), based upon large samples of national newspapers in the U.S., on support for the Equal Rights Amendment, concludes that there is no support for the hypothesis that newspapers bias letter opinion and asserts that "letter writers share the opinions of their neighbors who do not write letters" (p. 391).

It is valid to question whether letters contributed by particular interest groups, formed to argue for one side of an issue, should be discounted as unrepresentative. For example, of the anti-2,4-D letters, approximately 25% were sent by officers of the South Okanagan Environmental Coalition (SOEC). If these are discounted, the percentage of letter writers favoring the use of 2,4-D increases to over 30%. On the other hand it could be argued that other potential letter writers did not contribute letters because they relied upon SOEC to voice their point of view. Also, the possibility cannot be discounted that some other less formal public group was seeking to influence opinion in the pro-2,4-D direction through letter writing. Proponents of the latter point of view did occasionally identify themselves as spokespersons for the 'silent majority' (*Kelowna Daily Courier*, 16 August, 1978, p.4).

The chronological variation in the data presented in Figure 7.3 is also worthy of comment. The very large number of letters contributed in 1978, almost half the total for the eight year period, may have been provoked by several events. By 1978 the environmental group formed to combat the use of 2,4-D had become well-organized and cohesive. They published a literature review of the effects of phenoxy herbicides that received wide distribution. Respondents to the mailed questionnaire indicated that this report had been more influential on their view of the problem than the copious numbers of APMP information booklets produced. The above review concluded that "there is a considerable body of evidence which indicates that the phenoxy herbicides in general, and 2,4-D specifically, pose a substantial threat to environmental and thus human health " ... Extensive research on the effects of 2,4-D on test animals indicates that the herbicide is teratogenic, carcinogenic and very likely mutagenic" (Warnock and Lewis, 1978).

A further opportunity for the environmentalists to enlist public support was created in 1978 by a legislative change that allowed public appeals against permits for 2,4-D application. The anti-2,4-D forces brought experts from across the continent to testify at the hearings which were widely reported in the media. This coincided with the height of publicity against the use of 2,4,5-T in Vietnam and helped reinforce an unfavorable attitude toward the use of chemicals in the environment.

Following 1978 the number of letters dropped dramatically. It is, however, noteworthy that the number of letters in favor of 2,4-D in 1981 was higher than the number against. One explanation is that throughout the 1970's it appeared as though 2,4-D was going to become the dominant method of control. As time progressed it became apparent that the anti-2,4-D forces were gaining the upper hand, preventing the use of 2,4-D. Those in favor of using the herbicide therefore began to write letters to protest against what was occurring. In other words, letter writers were prompted to react *against* the perceived dominant situation.

CONCLUSION

Given the overwhelming dominance of anti-2,4-D letters, it would be unwise to ignore them as an indication of the strength of opinion. Whether they are representative of the broader public is impossible to judge; however, the mailed questionnaire results suggest that the letters implied a stronger anti-2,4-D bias than existed in the population. This would support the views of some other researchers who have found letter writers to be more negative than the general public (Forsythe, 1950; Grey and Brown, 1970; Renfro, 1979). A letter analysis may have provided a more realistic impression of the strength of effective public opinion on the use of 2,4-D for the APMP, but would not have constituted grounds, on this basis alone, to terminate the program.

ACKNOWLEDGMENTS

The author would like to thank the Social Science and Humanities Research Council of Canada for financial support for this research and the International Division of the Council for providing funds to attend the Chiang Mai Conference.

REFERENCES

Buell, E.H. Jr. (1975). Eccentrics or gladiators? People who write about politics in letters to the editor. *Social Science Quarterly* 56: 440-449.

Coffey, B.T. and McNabb, C.D. (1974). Eurasian water milfoil in Michigan. *The Michigan Botanist* 13: 159-165.

Dearden, P.(1983). Anatomy of a biological hazard: *Myriophyllum spicatum* L. in the Okanagan Basin, British Columbia. *Journal of Environmental Management* 17: 47-61.

Dearden, P. (1984), Public perception of a technological hazard: a case study of the use of 2,4-D to control Eurasian water milfoil in the Okanagan Valley. *Canadian Geographer* 28: 324-340.

Forsythe, S.A. (1950). An exploratory study of letters to the editor and their contributors. *Public Opinion Quarterly* 14: 143-144.

Grey, D.L. and Brown, T.R. (1970), Letters to the editor: hazy reflections of public opinion. *Journalism Quarterly* 47: 450-456.

Haskins, J.B. (1967). People who write letters. *Editor and Publisher* 100: 38.

Hill, D.B. (1981). Letter opinion on ERA: a test of the newspaper bias hypothesis. *Public Opinion Quarterly* 45: 384-392.

Kates, R.W. (1962). *Hazard and Choice Perceptions in Flood Plain Management*, Department of Geography, Research Paper No. 78. Chicago: The University of Chicago, p391.

Kates, R.W. (1978). *Risk Assessment of Environmental Hazard* New York: John Wiley and Sons.

Newroth, P.R. (1979) British Columbia aquatic plant management program. *Journal of Aquatic Plant Management* 17: 12-19.

O'Riordan, J. and Collins, M.P. (1974) *Water-based Recreation in the Okanagan Basin*, Victoria: Canada-British Columbia Okanagan Basin Agreement.

Renfro, P.C. (1979). Bias in selection of letters to the editor. *Journalism Quarterly* 56: 822-826.

Rosenau, J.N. (1974), *Citizenship Between Elections*, New York: Free Press.

Smith, G.E. (1971), Resume of studies and control of Eurasian water milfoil (*Myriophyllum spicatum*) in the Tennessee Valley from 1960 through 1969. *Hyacinth Control Journal* 9: 23-25.

Volgy, T.J., Krigbaum, M., Langan, M.K. and Mashier, V. (1977). Some of my best friends are letter writers: eccentrics and gladiators revisited. *Social Science Quarterly* 58: 321-327.

Warnock, J.W. and Lewis, J. (1978) *The Other Face of 2,4-D: A Citizens' Report*, Penticton, B.C.: South Okanagan Environmental Coalition.

Water Investigations Branch (1980). *Answers to Frequent Questions Concerning Control of the Eurasian Water Milfoil in British Columbia*. Victoria: Ministry of Environment.

White, G.F. (1945). *Human Adjustments to Floods: A Geographical Approach to the Flood Problem in the United States*. Department of Geography, Research baper No. 29. Chicago: University of Chicago.

8

Farmers' Perceptions Of Pesticides As A Cotton Crop Protection Strategy

Iqtidar Husain Zaidi

INTRODUCTION

This preliminary report describes some of the results of research to help understand the way cotton farmers in Pakistan perceive the use of pesticides to control insects damaging their crop.

Cotton, according to the Pakistan Census of Agriculture for 1980 occupies about 24% of the total crop acreage, and within the main cotton area in the provinces of Punjab and Sind it occupies about 27% of the total cropped area (Agricultural Census Organization (ACO), 1984). Cotton is popularly known in Pakistan as 'silver fiber' and its importance was recognized in 1948 by the establishment of the Central Cotton Committee. In view of its value to Pakistan's economy, cotton inputs must be carefully managed to optimize production.

PESTICIDE USE ON COTTON

At the official level, use of pesticides is regarded as an important strategy for managing cotton pests. However a majority of farmers in all categories of land tenure and farm size lack proper understanding of the use of pesticides. They possess little awareness of the hazards associated with careless use of pesticides. The failure of the cotton crop in 1983-84 provides evidence of the way improper use of pesticides could damage the crop. It was estimated that nearly half of the damage to cotton production in both Punjab and Sind provinces was caused by pests (Ministry of Food, Agriculture and Cooperatives (MFAC), 1984). The persistent rain was a cause of increased levels of pest infestation. Sales of pesticides in 1983-84 increased by 88% in Punjab, but this was insufficient to prevent crop losses. In many cases spray timing was inappropriate, and the number of sprays was inadequate. There were also instances where the wrong pesticide was used, for example, pesticides intended for sucking pests were used against bollworms.

As noted in MFAC (1984), the hazardous situation of 1983-84 may recur, because Provincial Governments have not made any arrangements to involve farmers in initiatives for taking plant protection measures. The need for such an arrangement is urgent particularly since the government withdrew the subsidy on pesticides in 1980. With the enforcement of this new policy, procurement and marketing of pesticides shifted to the private sector in Punjab and Sind. Editorials, articles and letters to the editor in national newspapers have also emphasized the ineptness of farmers as an unfortunate cause of the cotton crisis, either due to misguidance by pesticide distributors or because of farmers' ignorance in using pesticides.

These pest problems arose despite provisions of the Pest Ordinance that if farmers failed to undertake spraying of their crops during pest emergencies, the Provincial Governments would arrange field spraying and recover the cost as arrears of land revenue. In addition, to help small farmers, interest-free loans of Rs. 6,000 (approximately US$375) for the purchase of pesticides were available from the bank.

Policy makers seem to be shifting emphasis from the exclusive use of pesticides to integrated pest management (IPM) (MFAC, 1984). This policy is supported by positive results achieved on the experimental farm of the Cotton Research Institute (CRI) in controlling pink-bollworms with the help of a sex-pheromone. However, the idea of IPM has yet to be disseminated to farmers as an alternative to the exclusive use of pesticides and the majority of farmers lack a proper understanding of the concept and philosophy of IPM. Even under IPM, pesticides will be used when pest populations reach the economic injury level. A study of the way farmers perceive pesticides as a strategy for managing pest hazards is a useful basis for a more realistic crop protection policy for cotton.

METHOD

The Indus plain, comprising the provinces of Punjab and Sind covers over 99 percent of the area under cotton in Pakistan. The areas selected for study, given in Table 8.1, are those where the concentration on cotton was greater than the average for these Provinces (ACO, 1984).

The data presented here are derived principally from the Pakistan Census of Agriculture for 1980 (ACO, 1984), covering the period up to the date of the government withdrawal of pesticide subsidies (Zaidi, 1984). This includes information on number and area of farms by size of farm, tenure classification, farm fragmentation, land utilization, irrigation, intensity of cropping, cropped area under various crops by size of farm, use of manures, fertilizers and pesticides (including plant protection measures by size of farm and tenure), indebtedness and investment in agricultural machinery and livestock ownership. The data are of variable quality. An important source of error arises from the definition of a farm as the aggregate area of land operated by members of one household, with or without the assistance of members of other households, regardless of location, size or title. Farms wholly uncultivated during the census year are also included. Thus, even lands in other villages or districts, whether continuous or not, are included in the farm area of the person who operates them

(Zaidi, 1967).

Information on education levels was obtained from the Population Census of Pakistan, 1980. In measuring the correlation between the use of plant protection measures and education, the latter is defined in terms of literacy, i.e. ability to read and write in the local language.

A statistical analysis of these data has been done to assess whether: (a) the cotton acreage covered by plant protection is associated with the acreage under the more rewarding upland or American cotton; b) educated farmers use more pesticide; c) pesticide use is associated with the income per capita (data on cash crop value taken from Pasha and Hasan (1982)); d) pesticide use varies with land tenure type; and e) pesticide use varies with farm size. These hypotheses are based on the assumption that by 1980 the farmers must have become aware of the advantages of the pesticides. Relationships a), b) and c) were tested using the Spearman coefficient of rank corelation; d) and e) were tested using the chi-square test.

RESULTS

In the study area, with the exception of Faisalabad, cotton is one of the three major crops. However, the acreage under cotton was not significantly correlated to the acreage covered by plant protection (Table 8.1). It is notable that, in many areas, the proportion of cotton under crop protection was very small. The rank correlation coefficients between both literacy and per capita income and the cotton crop acreage covered by plant protection were insignificant at the five percent level. These findings are supported by a micro-level study of a village in Sind (Zaidi, 1984).

There was a significant correlation at the five percent level between the acreage under American cotton and the crop area covered by plant protection (R = 0.45).

The use of pesticides in different districts did generally vary significantly with farm size and with land tenure (Table 8.2).

DISCUSSION

The withdrawal of pesticide subsidies in 1980 was based on an assumption by policy makers that farmers had developed sufficient awareness of the advantages and disadvantages of pesticides. The results of this study suggest that farmers in the study area are, to some extent, aware of the advantages of using pesticides in the cultivation of American cotton. However, further study of the detailed nature of this awareness would require more field studies. In some districts, such as Bahawalpur and Multan, in-depth interviews have shown that farmers have begun to care less for their winter crops on the expectation of a better cotton crop due to the investment of labor and other inputs including pesticides. They expect the cost of these investments to be outstripped by the value of the crop. This tendency has made farmers more vulnerable to the risks of pest damage on

TABLE 8.1
Data on cotton production and crop protection in the regions studied

District	Total cotton acreage	% of total cropped area under cotton	% of cotton area under American cotton	% cotton area under plant protection
Jhang	236,774	14.7	93.6	03.0
Faisalabad	168,960	8.2	57.2	04.0
Multan	886,834	30.1	88.6	16.0
Vehari	360,411	31.4	88.1	29.0
Sahiwal	520,215	18.1	77.2	22.0
Muzaffargarh	225,435	12.2	81.1	03.0
D.G. Khan	169,025	12.8	95.2	01.0
Bahawalpur	271,476	26.5	81.9	24.0
Bahalnagar	321,075	22.4	24.0	00.5
Rahimyar Khan	481,383	30.6	94.7	02.0
Hyderabad	211,785	26.9	94.2	24.0
Tharparkar	236,706	13.0	94.7	38.0
Sanghar	353,093	41.8	97.3	29.0
Sukkur	237,953	32.5	70.0	02.0
Khairpur	244,363	34.0	88.6	04.0
Nawabshah	308,016	30.2	87.3	02.0

Source: Pakistan Census of Agriculture, 1980.

cotton, as happened in 1983 when the cotton crop was damaged by heavy rains and pest infestation. The perceived loss potential was considerably enhanced.

As shown above, up to 1980, it was the large and medium farm holders, owners and owner-cum-tenants who generally benefited from pesticide sprays. These farmers were the socially and economically more influential, and small farmers, who were and are still subsistence farmers, could not afford to protect their crops from pests, even with subsidized pesticides.

TABLE 8.2

The relationships between the use of plant protection measures on cotton and farm size and land tenure (chi-square values for each district)

District	Farm Size	Land Tenure
Jhang	10.4	131.0
Faisalabad	105.0	148.0
Multan	11.5	491.8
Vehari	9.4	28.7
Sahiwal	58.6	1,503.4
Muzaffargarh	39.7	29.1
D.G. Khan	21.1	0.8
Bahawalpur	46.4	30.0
Bahawalnagar	43.5	41.0
Rahimyar Khan	280.9	8.3
Hyderabad	64.3	3,217.6
Tharparkar	22.2	199.3
Sanghar	28.7	24.9
Sukkar	24.4	0.0
Khairpur	67.3	125.9
Nawabshah	101.2	83.9
Critical value (at 5% level of significance)	9.5 (df=4)	6.0 (df=2)

REFERENCES

Agricultural Census Organization (ACO) (1984). Pakistan Census of Agriculture, 1980. Statistics Division, Government of Pakistan, Lahore.

Ministry of Food, Agriculture and Cooperatives (MFAC) (1984). Report on 1983-84 Cotton Crop and Future Strategy. Government of Pakistan, Islamabad.

Pasha, H.A. and Hassan, T. (1982). Development ranking of the districts of Pakistan. *Pakistan Journal of Applied Economics* 1 (2) : 157-192.

Zaidi, I.H. (1967). The spatial pattern of farm sizes in the Punjab region of West Pakistan. *Pakistan Geographical Review* 22: 61-76.

Zaidi, I.H. (1984). Farmers' perception and management of pest hazards: a pilot study of a Punjab village in the lower Indus Region. *Insect Science and its Application* 5: 187-201.

9

The Pesticide Dilemma In Malaysia

V. C. Mohan

INTRODUCTION

In Malaysia, pesticides are an integral part of most agricultural practices, being used on rubber estates, oil palm plantations, vegetable farms, padi fields, fruit orchards, forests and even some backyard gardens. They are also widely used in public health programs, such as fogging for dengue fever and the malaria and filariasis eradication programs. Pesticides and fertilizers have become an accepted part of the lives of thousands of farmers, estate workers and sprayers.

The economy is prospering as a result of the increase in agricultural productivity. However, workers exposed to pesticides often lack the necessary safety information and they are exposed to the hazards associated with pesticide overuse. A survey conducted in 1981 by Friends of the Earth in Malaysia (Sahabat Alam Malaysia (SAM)) revealed that at least 14 pesticides which are either restricted or banned in several countries are widely used in Malaysia. These include aldrin, chlordane, 2,4-D, DDT, dichlorvos, dieldrin, endosulfan, endrin, HCH (BHC, 'Lindane'), heptachlor, leptophos, paraquat and 2,4,5-T.

There are several scientific studies on the effects of pesticides on human health in Malaysia, as well as local literature on field studies and research into relevant legislation.

Environmental groups and concerned individuals have, for the past decade, attempted to convince the Ministry of Agriculture of the dangers associated with pesticide use in Malaysia. This paper discusses some of the major problems and the role of non-government organizations in helping to resolve them.

ROLE OF THE GOVERNMENT

The Pesticides Board, the sole authority responsible for regulating pesticide use in Malaysia, is a sub-division of the Ministry of Agriculture and came into existence under the Pesticides Act, 1974. It took two years for the Board to draft rules on the registration of pesticides and another five years before these rules came into force in April, 1981. Only pesticides registered by the Pesticides Board are allowed to be manufactured, sold or used in Malaysia. However, as noted above the Board has registered as safe for use several pesticides that are either banned or restricted in many other countries.

The role of the government in Malaysia has been to stimulate the use of pesticides in general. The following extract from De Ashworth and Balasubramaniam (1975) exemplifies their point of view:

> In the long term the potential for expansion of the pesticide market in Malaysia would seem excellent as the Government is planning considerable increases in acreages of crops and is encouraging crops such as groundnuts, cocoa, maize and soya bean. This is associated with a relatively high wage level and the economic need for increased production suggests that farmers will become more dependent on herbicides and other crop protection chemicals. There is a happy relationship between Government and Industry while the country's central position geographically in South East Asia could make Malaysia very suitable for the setting up of additional pesticide manufacturing or formulation plants with supporting field stations.

International aid organizations have also supported the adoption of green revolution techniques involving fertilizer-responsive, pesticide dependent, high yielding varieties of food and cash crops.

The government's 'happy relationship' with the agrochemical industry extends to allowing firms to advertise their products in the Journal of the Ministry of Agriculture. Pesticide promotion is also undertaken by the government, and under the Farmers Organization Authority it has about 200 retail shops throughout the country, one of whose functions is to distribute pesticides. Pesticides are also distributed free of charge on specific government projects. The official attitude to the regulation of pesticides has, as perceived by SAM, given rise to several problems (Rajeswari Kanniah, 1983).

The law requires pesticide manufacturers to submit toxicological and other health and safety research data which are then 'critically reviewed' by the Pesticide Board before a product is registered. However, Malaysia does not have the scientific expertise, facilities or resources, to evaluate critically the data furnished, and the Board has to rely heavily on health and safety information provided by chemical companies. For example, in the case of 2,4,5-T, the Secretary of the Pesticides Board has admitted that Malaysia has no independent testing facilities to verify data submitted by chemical companies (De Ashworth and Balasubramaniam, 1975).

The dangers of such a situation became clear when the International Bio Test (IBT) Laboratories, the largest independent research laboratory in the USA, was discredited for deliberately falsifying toxicological data on pesticides. Subsequently, in the USA, the National Coalition Against the Misuse of Pesticides (NCAMP) prepared a list of 212 pesticides that had gained certification using suspect IBT data. SAM drew the attention of the Pesticides Board to the fact that 58 of the pesticides registered in Malaysia appear on the NCAMP list. However, it transpired that the Pesticides Board had been aware of the IBT scandal for over two years, but felt that its other sources of information were sufficient for its purposes. Nevertheless, the Board has admitted that it lacks the expertise and equipment to undertake studies on its own behalf and that it has not undertaken any toxicological studies since it was set up ten years ago (Star Publications, National Daily Paper, 26-2-1984).

Another pesticide problem in Malaysia has been the revelation that Britain sprayed areas around Bentong in Pahang with the herbicide 2,4,5-T, containing the contaminant dioxin, during its campaign against communist guerillas in the early 1950s. SAM has questioned whether Malaysians were employed to do the spraying, when and where spraying was carried out, and where the empty containers were disposed of, but the Pesticides Board and other concerned authorities have not initiated any inquiry to allay the fears of the Malaysian public.

Under these circumstances, SAM has called for the Pesticides Board to be replaced by a full department with *effective* powers to regulate and control the pesticide industry in Malaysia. While much is done actively to promote the use of pesticides, little is done to educate farmers or factory workers on the safe use and handling of pesticides.

PESTICIDE PROBLEMS IN MALAYSIA

Pesticide Usage and Poisoning

Malaysian farmers and workers usually apply pesticides using a back-pack sprayer, clad only in T-shirts, shorts and slippers. When powered sprayers and fogging machines are used, the only protection against inhalation of toxic fumes is a handkerchief or towel to cover the mouth and nose. These offer hardly any protection.

In 1980 SAM conducted a survey of pesticide use among farmers and estate workers in the state of Penang. The survey revealed that the majority of farmers were using pesticides every four or five days as a preventive measure. Most of them used pesticides up to nine days before harvesting and in some cases they were used on vegetable plots up to two days before harvesting. A large majority of the farmers changed their pesticide brands once every two or three years as they found the new brands more effective. Most did not follow the warnings or instructions given on pesticide labels.

Many padi farmers and small holders depend on their neighborhood shopkeeper for advice on the type of pesticide to use and how to use it. Shopkeepers in rural areas stock pesticides along with food and other items, and one can buy pesticides such as 2,4,5-T and paraquat in old syrup, soya sauce or soft drink bottles. In the Cameron Highlands, vegetable farmers are often given free samples of new pesticides in the market.

In Sarawak (East Malaysia) 70% of the farmers use pesticides. The majority of these farmers are illiterate, have little knowledge of the safe use of pesticides and cannot follow instructions on labels. Farms are located in the interior and are relatively inaccessible, except for river transport during favorable weather conditions. In such circumstances there is little effort by agriculture officers to educate farmers on the proper use of pesticides and the officers are not available in an emergency.

In a major outbreak of brown plant hopper on rice in Kedah in 1980, proper safety precautions were not taken during an extensive pesticide spraying exercise. This resulted in 30 farmers being hospitalized after exposure to toxic chemicals and one death. In 1983 rice farmers in several Malaysian states complained of a mysterious disease affecting fish in their fields, causing sores (*wabak kudis*), and farmers were advised by agriculture officers not to eat the fish. A committee set up by the Ministry of Agriculture concluded that pesticides were the cause of the fish disease. It appeared that padi field fish had developed some degree of tolerance to pesticides and, although the chemicals no longer killed them, they caused the development of sores on their bodies, due to hemorrhaging septicaemia (SAM, 1981). Research has also indicated that, in pesticide manufacturing and repacking industries in Malaysia, there is little effective protection against toxic chemicals for factory workers:

- At pesticide factories in the Klang Valley area, workers are not provided with adequate protective clothing. In one factory, workers were not provided with face masks or respirators when working with powder and volatile liquid chemicals. They are, however, provided with gloves.

- Workers from a factory which supplies the fungicide thiram (wettable powder) complained of body itch, even from a small amount of powder on their skin.

- The liquid herbicide dalapon (Dowpon) is packed by workers at a company in Shar Alam. They are provided with face masks but they say the masks are not effective as they can still smell the chemical and suffer from dizziness after long hours at work.

- At Wesco Chemicals in Kepong, where pesticides imported in bulk are repacked, workers are not provided with uniforms, safety shoes, gloves or face masks.

In a major survey of plantation workers on 30 rubber and oil palm estates, SAM found a large number of workers suffering from the after-effects of prolonged exposure to pesticides. In Glenmarie Estate in Selangor, 20 women workers were suffering from skin disease and nose bleeds attributed to constant exposure to weed-killers and fertilizers. Nine were badly affected, with toe and finger nails dropping off.

In big plantations, management decides on the nature, dosage and frequency of application of pesticides. Plantation workers are illiterate and ill-informed and often become victims of a system which places more emphasis on profit than on their health and well-being. The following deficiencies were found in the survey:

- No protective clothing whatsoever is provided.
- Workers who spray herbicides suffer from skin irritation and other skin diseases.
- Laborers are not given proper medical care, being examined only once a year or not at all.
- No adequate facilities are provided for workers handling herbicides to wash themselves before eating.
- No training or information is provided to laborers on the dangers posed by the chemicals used.
- Herbicide-contaminated clothing is washed together with the family's other clothes.
- Workers use empty pesticide containers and bottles to store water, cooking oil and other consumables.
- On some plantations, the management instructs its workers to take care of their own spraying pumps and workers normally take the pumps home where they may be within the easy reach of children.
- On most estates spraying is done by girls and women, and in some cases by pregnant women. A spray pump normally contains four gallons of water and in one day they will carry at least 20 loads of herbicide.
- On many estates dangerous chemicals are kept close to food.

There is evidence that paraquat poisoning is becoming unacceptably common in Malaysia. In a study of 30 cases in the Kuala Lumpur General Hospital between 1978 and 1979, there were 27 deaths and three survivors (Chan and Cheong Izham, 1982). Deaths occurred from five hours to 22 days after ingestion, with a mean survival time of five days. Indians, who comprise the main workforce on the estates, were the predominant racial group in paraquat poisoning cases (67%). The report called for the dangerous habit of decanting paraquat into unlabeled, common household bottles to be made illegal by law.

The Deputy Agriculture Minister has stated that paraquat need not be banned because it is cheap and effective. However, the Health Ministry has stated that it is concerned about the use of paraquat as there have been more than 300 cases involving paraquat poisoning, of which more than a third were fatal.

Wong Kieng Keong (1981) has shown that the levels of pesticide residues in the blood of the general population of Malaysia are much higher than in the general population in the United States of America.

Another study (Choy, 1981) revealed that Malaysians could have three to eight times more DDT in their blood than Americans and that DDT levels in the blood serum of spraymen in the Malaria Eradication Program is between six and ten times higher than that found in other Malaysians (whereas levels of organochlorine insecticides were similar in spraymen and the general population, suggesting that both have similar sources of intake of other organochlorine insecticide residues, presumably from food). Twelve organochlorine insecticides were identified and quantified, including HCH, aldrin, dieldrin, heptachlor epoxide and DDT. These findings support earlier comments on the extent of pesticide pollution in Peninsular Malaysia and the weakness of present controls on the use of organochlorine insecticides.

Many government officials appear unaware of the pesticide problem. The Secretary General of the Ministry of Agriculture, recently recommended that farmers should employ preventive measures such as spraying three to four times during the life cycle of a crop with different chemicals each time. Such an approach is contrary to the principles of integrated pest management (which the Ministry of Agriculture professes to support) and may promote the development of pest resistance to pesticides.

Pesticides and Pest Resistance

Indiscriminate spraying of pesticides has favored the selection of resistant strains of pests and the outbreak of plant pests on an epidemic scale. This is a relatively new phenomenon which emerged largely after 1976. Over the past few years, rice crops have suffered the following pest infestations:

- In July 1977, a brown plant hopper (BPH) (*Nilaparvata lugens*) epidemic in Tanjung Karang, Selangor caused the destruction of 8,000 ha of rice ready for harvesting. A further 648 ha was burnt to prevent BPH from migrating to neighboring fields.

- In January 1978, insecticides failed to stop BPH from destroying more than 40 ha of padi fields in Southern Kedah.

- In June 1979, despite widespread use of pesticides, BPH destroyed $2 million worth of crops in Kedah.

- $5 million worth of rice was lost during the 1981/82 growing season due to the Malayan black rice bug (*Scotinophara coarctata*), and another $3 million worth due to the tungro disease known as *Penyakit Merah*.

- In 1982, 39,000 ha of rice were attacked by eight major pests and diseases, including BPH, white-backed plant hoppers, tungro, rats, leaf folders, black rice bugs, rice stem borers, rice grain suckers and blasts. Losses were estimated at $31 million, and tungro alone accounted for losses of $24 million. Agricultural officers advised farmers to burn affected fields.

- In 1983, the same pests attacked 38,000 ha of padi land, causing an estimated loss of $23.2 million, $14 million of which was due to tungro.

- By 1983, padi planters in the Muda agricultural region were expected to lose $25 million every season if the tungro disease is unchecked.

In 1984 agricultural officers advised farmers to stop irrigating their fields so that the disease-carrying green hoppers (*Nephotettix apicalis*) would be deprived of breeding grounds. Muda district authorities who had previously combated pest infestations by the application of pesticides, stated that the use of chemicals on the scale required would be unsuitable, because of potential side-effects on people.

In public health, pesticides such as dieldrin, HCH, DDT, malathion and pirimiphos methyl have figured prominently in malaria and filariasis eradication programs. According to an Institute for Medical Research report, two strains of mosquito have developed resistance to dieldrin and HCH, one as a result of household spraying of dieldrin against adult mosquitoes. In this strian, resistance to dieldrin was about 100 times more than that prior to treatment in both adults and larvae; resistance to HCH in larvae was about 20 times more, while resistance to DDT was slight.

A recent WHO report states that 51 species of anopheline mosquito have developed resistance to one or more insecticides; 34 are resistant to DDT; 47 to dieldrin; 30 to both DDT and dieldrin; 10 to organophosphate insecticides.

There is further evidence that several common Malaysian pests are beginning to develop multiple and cross resistance to a variety of pesticides. For instance, in the Cameron Highlands the diamond back moth *Plutella xylostella*, has developed resistance to both organophosphates and carbamates (although only the former have been widely used). The development of resistance has prompted farmers to use extremely toxic unregistered pesticides which have been smuggled into the country or to mix pesticides to boost their effectiveness.

It is regrettable that the situation has been allowed to deteriorate to the point where Malaysian farmers are suffering severe crop losses and also endangering their health and safety by the excessive use of pesticides.

REFERENCES

Chan, K.W. and Cheong Izham K.S. (1982) Paraquat poisoning: a clinical and epidemiological review of 30 cases. *Medical Journal of Malaysia* Vol. 37 No. 3.

Choy, C.K. (1981). A study on organochlorine insecticide residues in the blood sera of Malaria Eradication Program Spraymen and members of general population in Peninsular Malaysia. Universiti Pertanian Malaysia, p. ix.

De Ashworth, B. and Balasubramaniam, A. (1975). Crop protection in Malaysia. Paper presented at Collaborative Pesticides Analytical Council's Symposium at Oerias, Portugal, June 1975.

Rajeswari Kanniah (1983). Pesticides abuse in Malaysia — problems and issues.   Paper presented at S.M. Seminar on Problems of Development, Environment and the Natural Crisis in Asia and Pacific, Penang, 22-25 October 1983.

Sahabat Alam Malaysia (1981). *Pesticide Problems in a Developing Country — a Case Study of Malaysia.* — SAM.

Wong Kieng Keong (1981). Impact of pesticide usage — a case study of organochlorine compound levels in the blood serum of selected Malaysian population groups. CAP Seminar on Economics, Development and the Consumer, 1980. *World Health Organization Chronicle* Vol. 35 pp 143-148.

10

Marketing Of Pesticides In Pakistan In Relation To Legal And Other Controls

Mushtaq Ahmad

INTRODUCTION

In Pakistan more than 70% of the population derives its livelihood directly or indirectly from agriculture. The crop area is about 40 million acres (Ma), the major crops being wheat (18.2 Ma), cotton (5.4 Ma), rice (4.9 Ma), grams (2.3 Ma), sugarcane (2.2 Ma), maize (1.9 Ma) and fruit (0.9 Ma). More than 60% of the pesticides used are applied to cotton, about 10% to rice and the remainder is primarily used on sugarcane, fruit, vegetables and grain legumes.

In Pakistan, prior to January 1980, pesticides were marketed almost entirely in the public sector. After importing pesticides, the government handed over a small portion to the private sector for sale to farmers. Most were sold to farmers by the government itself through its own outlets. There were heavy subsidies on pesticides, up to 75% on granules and 50% on emulsifiable concentrates, dusts and wettable powders. Free aerial spraying was organized, particularly on cotton and sugarcane crops, by the central government using its own aircraft. In order to streamline plant protection in the country the government withdrew the subsidy in February 1980 and transferred the importing, distribution and marketing of agricultural pesticides to the private sector who welcomed this change and immediately set up their own distribution network. The consequence of this has been a considerable increase in the use of pesticides as shown in Table 10.1 (Anon, 1985). Between 1981 and 1984, the money value of pesticide sales increased five-fold while the quantity of active ingredient only trebled. The government has retained control of the import and use of pesticides in public health.

The insecticide market is presently dominated by the pyrethroids which constituted about 45% of sales in 1984. Organophosphorus insecticides captured 39% of the market, carbamates 4% and chlorinated hydrocarbons 9% during 1984. Apart from small quantities of HCH and DDT, which are manufactured locally, all pesticides are imported, mainly from Switzerland, Germany, the Netherlands, Japan, USA and Italy.

TABLE 10.1

The increase in pesticide use from 1981-1984

Year	Quantity of pesticide used (tons a.i.)	Market value (million Rupees†)
1981	905	225
1982	1320	345
1983	1756	626
1984	2600	1200

† Current exchange rate: 16 Rupees to the US Dollar.

THE MARKETING SYSTEM FOR PESTICIDES

The marketing of pesticides is in the private sector in Pakistan, except for one province, Baluchistan, where pesticide use is less than 2% of the total national consumption. This province is mostly desert and hilly, with little cultivation in the scattered communities. Fearing that the private sector could not supply such a distant area because of the small amount of business, the provincial government imports and sells pesticides to farmers through its own outlets. In the remaining three provinces, Punjab, Sind and North West Frontier Province, the private sector plays a full role in making pesticides available to farmers. There are approximately 34 distributors in these provinces.

Each distributor has a network of dealers in all the provinces, located in strategic areas. The distributors have established storage points in different parts of the country which supply pesticides to the dealers. Each distributor has technical staff to give advice to dealers and farmers on the correct use of pesticides. Technical information leaflets on the correct use of pesticides and their dosage rates for different crops are provided free in the local language. Pesticide promotion to increase the use of pesticides, is organized through television, radio, newspapers, demonstration plots and farmers' meetings.

LEGAL AND OTHER CONTROLS

Pesticides are controlled in Pakistan by various restrictions, advice and laws issued by the government regarding registration, import and sale of pesticides.

Registration

The foremost control on pesticide marketing is that no pesticide can be sold unless it is registered under the Agricultural Pesticides Ordinance

(Anon, 1971) which states that "No person shall import, manufacture, formulate, sell, offer for sale, hold in stock for sale, or in any manner advertise any brand of pesticide which has not been registered in the manner herein after provided." Procedures are laid down in the ordinance for registering a pesticide, and the conditions to be fulfilled are specified on the application form. These conditions are similar to those in other developing countries, but not as stringent as in western countries. If one fulfills the following legally-based conditions, a pesticide can be registered within two years:

i) the brand is not such as would tend to deceive or mislead the purchaser with respect to the guarantees relating to the pesticide or its ingredients or the methods of its preparation;

ii) the guarantee relating to the pesticide or its ingredients is not the same as that of any other registered brand by the same manufacturer or is not so similar thereto as to be likely to deceive;

iii) it is effective for the purpose for which it is sold or represented to be effective;

iv) it is not generally detrimental or injurious to vegetation, except weeds, or to human or animal health, when applied according to directions.

In order to satisfy themselves that the above conditions are met in letter and spirit, the government have specified procedures in the Ordinance. As on 31 December 1984, 147 pesticides have been granted registration, 57% being insecticides, 18% fungicides and 14% herbicides.

Label Requirements

In Pakistan, most emphasis is put on meeting increasingly detailed label requirements; by 1984, the following requirements had to be met on every pack of pesticide for sale: name of product; name and address of manufacturer or formulator or the person in whose name the pesticide is registered; net contents; date of manufacture; registration number; date of test; normal storage stability; name and percentage by weight of active ingredient and total percentage by weight of other ingredients; the words "keep out of reach of children"; the word "DANGER"; the word "POISON" in red on a contrasting background; a picture of the skull and crossbones; a statement on antidotes; direction to call a physician immediately in case of poisoning; directions for use; directions to destroy empty containers and bury them in the ground; expiry date; retail price; batch number; any other information useful for farmers such as dosage per acre or per unit of water for different crops, plants etc.; the words "only for agricultural use"; circle color code according to the LD_{50} and one of the following four explanations, as applicable: a) Extremely Toxic (Restricted); b) Highly Toxic; c) Moderately Toxic or d) Toxic.

Label Integrity

The Registration Authority of Pakistan have coined the phrase 'label integrity' and have circulated the following directions on pesticide labeling:

i)   the label should not describe a product by such terms as 'safe', 'harmless', 'non-toxic', 'non-poisonous' or 'non-injurious' in respect of risks to humans or animals, either with or without such qualifying phrases as 'when used as directed';

ii)   there should be no use of superlatives concerning a product, for example 'the best', 'most effective', or 'superior control' or 'unrivaled';

iii) the technical information on recommendations for use of a product should be clear and specific;

iv)   there should be no false, unjustifiable or potentially misleading statements on the label as to the name, origins, constituent, composition, effectiveness, safety or other attributes of the product;

v)   practical advice should be included on methods of preparing and using the product, for example opening, measuring, mixing, agitation;

vi)   warning should be given, where necessary, of the time interval which must be allowed before sowing or planting a repeat or following crop;

vii) any special recommendations on storage conditions for the containers and product should be included.

Control of Pesticide Imports

Even registered pesticides are not freely importable. All provinces have set up expert committees consisting of government officials and representatives of trade and industry to scrutinize the indents of distributors and make recommendations to the Chief Controller of Imports for the granting of an import license, on the basis of which a pesticide can be imported into the country.

A distributor has to submit his requirements for a registered pesticide to the Expert Committee of the respective province, supply its recommendations to the Chief Controller of Imports, and include the following information: name of pesticide; formula of the pesticide; packing; name of manufacturer; quantity; source of supply; photocopy of the certificate of registration; retail selling price; a copy of the label.

Similar requirements must be met for the import of active ingredients of a pesticide for local formulation within Pakistan. It is evident from the above that there should be no secrecy about the quantities of pesticides imported, their import price or source of supply.

Sales Reports

All distributors are required to submit quarterly sales reports to the province authority, giving the following information: name of pesticide; unit; opening balance, quantity and value; fresh arrivals, quantity and value; grand total, quantity and value; quantity sold, quantity and value; balance on close of the quarter, quantity and value.

Submission of Indents for Import

Similarly, the following information is required regarding the utilization of import licenses: date on which the indent was submitted to the provincial government; recommendations of the provincial government; date on which the firm applied for the license; grant of import license by Chief Controller of Imports and Exports; date of license; shipment schedule.

Sale

There are many government restrictions and controls on the sale of pesticides. Every distributor has to be registered with the Provincial Agriculture Department before being allowed to market pesticides in the province. The following information has to be provided in the application: name and address of firm; whether the firm is Public Limited or Private Limited; photocopy of registration under Companies Act and income tax registration number; registered head office; foreign and/or Pakistani investment shares; capital; annual turnover; bankers and auditors; type of business and commodities which the firm deals in at present; experience in pesticide marketing and pest control operations in the country; number and names of technical staff together with their qualifications and experience; names, qualifications and experience of distribution agents and particulars of sales depots in each district of the country; number and type of plant protection equipment available; pesticide formulation and packing facilities; storage facilities for pesticides with full particulars; transport facilities for pest control and other agricultural work in the country; whether already registered as pesticide distributor in any other province and when the present agreement is to expire.

According to the latest instructions from the government, dealers in pesticides are also required to be registered with the provincial governments. If a dealer is selling pesticides from more than one distributor, a separate registration is required for each distributor. The provincial governments are insisting that distributors of pesticides must have some sales points in the less accessible areas in the arid zone of the country, where few crops are cultivated, on a limited acreage and hence demand for pesticides is extremely small. Marketing pesticides in such areas is unprofitable.

Quality Control

The government has appointed inspectors who have power to check the quality of pesticides. An inspector may, within his local area, enter any premises where pesticides are kept or stored, whether in containers or in bulk, by or on behalf of the owner (including premises belonging to a bailee, such as a railway, shipping company or any other carrier) and may take samples for further examination. No compensation will be payable for a reasonable quantity taken as a sample. Where an inspector takes a sample of pesticide for testing or analysis, he must intimate this purpose in writing in the prescribed form to the person from whose possession he takes it. Also in the presence of this person, he must divide the sample into three portions, effectively seal and suitably mark the same, and permit him to add his own seal and mark to all or any of the portions.

Where the pesticide is made up in small volume containers, instead of dividing a sample as above, the inspector may take three containers, after suitably marking them and, where necessary, sealing them. This also applies if the pesticide is likely to deteriorate or be otherwise damaged by exposure.

The inspector should leave one portion of a sample so divided, or one container, as the case may be, with the person from whom he takes it. Of the other two, he will send one portion or container to the government analyst for testing or analysis, and the other portion or container to the central government.

The government analyst to whom a sample of any pesticide has been forwarded by an inspector will deliver to him a signed report of the result of the test or analysis.

Price Control

The government authorities have so far allowed a free hand to trade and industry when setting the price of pesticides to the user, but pressure has been mounting against the increase in prices over the last few years. The government's contention is that the mark-up on the price of a pesticide should not exceed 80% of its cost and freight (C&F) Karachi value. The Federal Pesticide Committee was constituted by the government in 1980 to implement the decisions of the cabinet regarding the new agricultural policy of transferring the marketing of pesticides from the public to the private sector and they have reviewed the price structure of all pesticides. For those pesticides whose increase in price exceeded the limit of 80% of C&F value, explanations and justifications were called for from the distributors concerned. The position was adequately explained and justified but the government do not appear to have been satisfied. Consequently the whole question of the sale price of pesticides has been referred to the Agricultural Price Commission which will review the position and report its recommendations regarding the appropriate price level of pesticides to the government,after which a final decision will be taken.

Environmental and Health Controls

In addition to the above mentioned controls and constraints on the marketing of pesticides, there are at times objections from environmentalists and those with an interest in wildlife, regarding excessive and improper use of pesticides, particularly in areas near fish ponds, beekeeping farms and wildlife parks. The problem is not yet acute but as a result of the Bhopal tragedy in India several social organizations, government departments and enlightened individuals have expressed fears about the ill-effects of pesticides. The press has also taken an increasing interest in this matter. Until now the government has not fixed residue tolerances for pesticides, nor is any license needed for applicators who handle extremely toxic pesticides, as is required in many western countries.

IMPLEMENTATION PROBLEMS

The controls on marketing of pesticides in Pakistan, particularly the label requirements, seem exhaustive and very useful on paper, but in actual practice these are not always implemented according to the letter and spirit in which they were constituted. The label requirements are often printed in English, which most users cannot read.

The registration procedure also takes a very long time. A pesticide is seldom registered in less than three years. Sometimes registration is delayed by petty objections by the Registration Authorities.

As quality control laws and regulations are not being properly implemented, the farmer is at the mercy of manufacturers with regard to the quality of the pesticides. On the other hand manufacturers have to spend a lot of time obtaining permission from the government for the import of pesticides to Pakistan.

REFERENCES

Anon (1971). *Agricultural Pesticides Ordinance*. The Government of Pakistan, Pakistan.

Anon (1985). *Statistical Year Book*. The Government of Pakistan, Bureau of Statistics, Pakistan.

11

Pesticide Flow And Government Attitude To Pests And Pesticides In Kwara State, Nigeria

Oluwayomi D. Atteh

INTRODUCTION

In the last few decades, efforts to increase food production in the third world have drastically changed traditional agriculture. Machinery, fertilizers, pesticides and improved varieties of crops have been introduced. The green revolution has spawned massive importation of pesticides, and there is a growing concern about the potential harmful effects of intensive use of these chemicals (Bull, 1982; Goldman, 1982; Napompeth, 1981). Experience in the U.S., Great Britain and other countries where pesticides have been used extensively should alert third world countries to their potential hazards.

Atteh (1984) described the perceptions of pests and pesticides of Nigerian farmers. The present paper is part of a national study and reports on the distribution and use of pesticides in Kwara State, Nigeria.

This paper describes sources, types and distribution of pesticides within Kwara State and examines government quality control policy and awareness of potentially harmful side-effects of pesticide use. Kwara, in west central Nigeria, was one of the 19 states in Nigeria created in 1976. It is predominantly guinea savanna in the 'middle belt', with forest to the south and open grassland to the north (Ireland, 1962).

The State occupies 60,388 sq km, 7% of the land area of Nigeria. According to the 1963 census it had a population of 1.7 million and it was estimated to have 2.5 million by 1977 (Ministry of Finance and Economic Development, 1978). It is divided into 12 Local Government Areas. Its capital is Ilorin (estimated population 350,000). Over 65% of the labour force was, until recently, engaged in agriculture with the major crops being yam, sorghum, cassava, maize, beans, melon, okra, pepper, soybean, potato, cocoa and coffee. Farming is typically done by small-scale peasant farmers whose farms average 1.7 ha and are fragmented into many plots. In the past 15 years, the oil boom has enhanced the non-agricultural sectors of the economy, drawing the labour force away from agriculture. Most people in the 15-35 year age group now live in urban centers, depleting the farm labour force (Atteh, 1980; 1984; de Vos, 1975).

In the past ten years, Kwara State Ministry of Agriculture and Natural Resources (SMANR) has mounted a strenuous campaign to encourage the use of chemical pesticides to increase crop yields (Atteh, 1984). Radio, television, cinema and pamphlets have been used in the campaign. Many traditional pest control methods employed by farmers in this area (Atteh, 1984) have been displaced by chemicals.

PESTICIDE PROVISION AND DISTRIBUTION IN KWARA STATE

Information was collected by interviewing the following: SMANR officials responsible for pesticide distribution and recommendations; agents of major pesticide manufacturers and retailers; and 120 farmers and heads of households selected at random. Interviews were conducted in 1983 and 1984.

Sources and Types of Pesticides

Several parallel systems of pesticide distribution exist, as indicated in Figure 11.1. Agricultural pesticides are distributed by both government and private sources. Kwara SMANR purchases pesticides or receives them through aid donors and then distributes them to farmers. Private firms also sell directly to farmers. Non-agricultural pesticides, used mainly against household pests, are usually provided through commercial channels. However, the Ministry of Health provides some pesticides for the control of public health pests.

The SMANR obtains agricultural pesticides from three primary sources:

Pesticide companies — Ciba-Geigy through its local subsidiary Swiss Nigeria Chemical Company Limited; a local subsidiary of ICI, Chemical and Allied Products (Nigeria) Limited; and Dizengoff. These companies have local agents and offices throughout Nigeria who advertise and sell their products and lobby influential government officials. Kwara SMANR deals with companies through its agents based in Ilorin.

Federal agency — the Pest Control Division of Nigeria's Federal MANR (FMANR) has agricultural pest control programs in all states. The FMANR purchases pesticides in bulk and distributes them to SMANRs annually or twice a year. The kinds and quantities of pesticides issued to each state depend on availability of materials, and outbreaks of specific pests. The FMANR issued the following pesticides to Kwara SMANR in 1983 and 1984: endosulfan, fenitrothion, pirimicarb, coumatetralyl, 'Pernithrothion', 'Metrab', trichlorfon, MCPA, ethiofencarb, monocrotophos, aluminum phosphide, phosphamidon, malathion, propoxur, thiram, alachlor, tetrachlorvinphos. The SMANR officials interviewed said they had little control over the kinds of pesticides the FMANR issues, cannot determine in advance

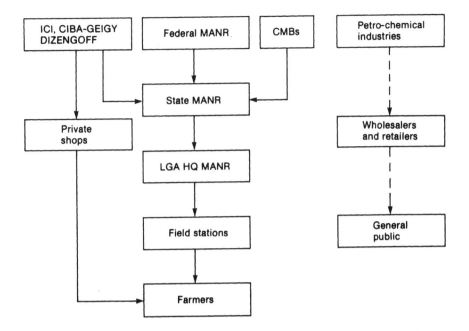

Figure 11.1   Pesticide distribution pattern in Kwara State

Key: ——► Flow of agricultural pesticide; — — ► Flow of non-agricultural
pesticide;   MANR = Ministry of Agricultural and Natural Resources;   CMBs = Commodity
Marketing Boards

when the materials will arrive and often receive no information concerning toxicity or application procedures for the pesticides.

Nigerian Commodity Marketing Boards (CMBs) — these Boards purchase
bulk quantities of various agricultural products, including pesticides,
which the SMANRs distribute to farmers. The SMANR has little control over what pesticides the CMBs issue or the time of issue and may
not receive information on the toxicity of the materials.

Recently, a few retail shops have opened, selling pesticides produced
by the three companies listed above. In July 1984, there were ten such
shops in Ilorin and 12 in eight of the remaining 11 Local Government Area
(LGA) headquarters. However, most farmers are in areas remote from
shops and rely on the SMANR for pesticides.

The SMANR generally restricts its pesticide purchases to the following
products (amounts purchased in 1982, if available, are shown in
parenthesis):

Herbicides — paraquat (34,019 litres), general purpose contact weed killer;
atrazine (52,480 litres), for use on cereals; prometryne (3,750 kg), for

use on yam, cassava and potatoes; thiobencarb, for use on rice; propanil, for use on rice.

Crop insecticides — dieldrin (Dieldrex 20) (1,463 litres), for use against termites; 'Kokotien' (1,141 litres), for use against grasshoppers; aldrin dust (5,000 kg), for use against termites, crickets and other soil insects affecting yam seed stock; 'Damfin' (2,000 kg); aluminum phosphide, for protection of stored cereals.

Pesticides for Livestock — the SMANR's Veterinary Division distributes the following major pesticides (data on amounts unavailable): coumaphos (Asuntol) and dioxathion (Delnav DFF) for control of ticks, lice and fleas; malathion; benzyl benzoate (Ascabiol); ivemacetin (Ivomec) for control of external and internal parasites; griseofulvin (Griseolfin) for control of ringworm.

In addition, the FMANR Pest Control Division applies large quantities of DDT (75% wettable powder) and dieldrin to vegetation along cattle trails to control tsetse fly, the vector for trypanosomiasis, and proved infested bushes. DDT and dieldrin, applied to pools of stagnant water, are also used in the campaign to eradicate malaria.

Because of its inadequate budget and a low demand from farmers, the SMANR did not directly purchase any pesticides in 1983 or 1984. In October 1984 more than 60% of the pesticides purchased in 1982 had still not been distributed.

Pesticide Distribution System

Figure 11.1 shows the system of pesticide distribution in Kwara State. Ninety five percent of those respondents who used pesticides listed the SMANR as their source, chemicals being given to farmers on request.

Each of the 12 LGA headquarters has an agricultural officer supervising the extension agents and field stations. The number of extension officers and field stations ranges from 15 in Oyun to over 50 in Ilorin and Oyi. Each extension officer serves about 1,000 farmers in Kwara State and most farmers have access to an extension agent. The agents serve as advisers on all agricultural matters including pests and pesticides. They recommend specific pesticides to combat pest problems and teach farmers how to use them.

Pesticides purchased by the SMANR are sold to farmers at subsidized prices (usually one third cost price), while those donated by the FMANR and CMBs are distributed free of charge to farmers. If a major pest outbreak occurs, the SMANR may intervene and apply pesticides free of charge. For example, outbreaks of rinderpest in Ilorin, Edu and Borgu LGAs in 1982 and 1983 and of the grasshopper *Zonocerus variegatus* in Oyi LGA in 1982 were considered state emergencies, with chemical control being provided free.

Quality Control and Awareness of Hazards

SMANR officials rely on information provided by pesticide labels and distributors to determine the ingredients of the materials issued by the FMANR. The SMANR is not equipped to conduct independent analyses. Inspection of SMANR stores, and comments from SMANR personnel, revealed that many stored pesticides had exceeded their expiry dates or the expiry dates were not given on the containers.

The study also revealed that many SMANR personnel are unaware of the hazards presented by pesticides. The local field personnel of SMANR were generally much more aware of the potential hazards than higher SMANR officials. The latter reported that they depend on the FMANR and Federal Ministry of Health to ensure that the pesticides received are of good quality and relatively safe to use.

Until recently SMANR provided overalls, boots, gloves and soap to those using dieldrin, DDT and certain other pesticides considered potentially hazardous. However, because of the recent financial crisis in Nigeria, these safety materials are no longer issued. SMANR field personnel report that, as a result, pesticide-related health problems, e.g. eye irritation, nose bleeding, stomach aches, diarhoea and skin disorders have increased among pesticide users.

Forty-six residents in the Ilorin area in 1983 were hospitalized as a result of mistakenly drinking or eating pesticides. In many villages some infections and diarhoea have been attributed to eating food that had been stored in used pesticide containers. Two children were hospitalized with serious burns in April 1984 in Kabba because they threw empty 'Sheltox' containers into a fire.

Non-Agricultural Pesticides

The study showed that pesticides were very popular among house-owners for the control of household pests such as mosquitoes, cockroaches, mice, ants and flies. Over 15 brands of household insecticide were available in local markets, the most common being 'Sheltox', 'Mobil', 'Gammalin 20', 'Detol' and DDT. They could be purchased in most shops and gasoline stations, and were also sold in street markets.

CONCLUSIONS

Kwara State does not effectively control the distribution and use of pesticides. Inferior quality pesticides and products banned in their countries of origin can therefore find their way easily into the state.

The majority of pesticide users in the state are illiterate peasant farmers who are unable to read pesticide labels for information such as recommended dosage, application procedures or safety precautions. Even those who can read may be using pesticides from containers without labels. Further, SMANR officials may not be qualified to advise on correct use and safety procedures.

To combat these deficiencies, Kwara State should develop pesticide regulatory and improvement programs that include laboratory facilities and properly trained staff to monitor pesticide quality and ensure safe practices. All pesticide labels and instructions should be given in the major local languages.

In addition, the state and university personnel should carry out educational programs, using farmer demonstrations, radio and newspapers to encourage safe use of pesticides. The training should also encourage integrated pest management strategies.

REFERENCES

Atteh, D.O. (1980). Resources and Decisions; Peasant Farmer Agricultural Management and Its Relevance for Rural Development Planning in Kwara State, Nigeria. Ph.D thesis, University of London, London.

Atteh, O.D. (1984). Nigerian farmers' perception of pests and pesticides. *Insect Science and Its Application* 5(3): 213-220.

Bull, D. (1982). *A Growing Problem: Pesticides and the Third World Poor* OXFAM: Oxford.

Goldman, A. (1982). Pesticide use, hazards and perception in Kenya. Paper presented at the 3rd PMPP meeting, Nairobi 21-25 June, 1982.

Ireland, A.W. (1962). The little dry season of Southern Nigeria. *The Nigerian Geographical Journal* 5(1):7-20.

Ministry of Finance and Economic Development (MFED) (1978). *Population Projections for Kwara State — 1963-1980.* MFED, Ilorin, Nigeria.

Napompeth, B. (1981). *Thailand National Profile on Pest Management and Related Problems.* Special Publication 4, National Biological Control Research Center, Kasetsart University/National Research Council, Bangkok, Thailand.

de Vos, A. (1975). *Africa, the Devastated Continent* W. Junk: Hague, Netherlands.

12

Instability In Agroecosystems Due To Pesticides

C.B.S.R. Sharma

INTRODUCTION

It is estimated that over four million chemicals exist, with a few thousand added annually, to which man and other biota are exposed in the workplace, as residues in food or as general environmental contaminants. Toxicity to genetic material is a potential effect of these chemicals (Crow, 1983) and may occur at the gene, chromosome and/or genome level, resulting in mutagenicity, clastogenicity or turbagenicity, together constituting genotoxicity. In man this leads to carcinogenesis, teratogenesis and heritable diseases, adding to the social burden (Figure 12.1).

It is proposed here that the genotoxicity of pesticides is a potential threat to the stability of agroecosystems causing mutations, directly or indirectly, gradually or suddenly, and such pollution by imperceptible mutations may alter the delicate balance of the ecosystem.

EVIDENCE FOR GENOTOXICITY

Kurinnyi and Pilinskaya (1974) estimated that there were 800 individual pesticides, only 17% of which had been investigated and, of these, 70% were genotoxic in one or more systems. Ridgeway et al. (1978) estimated that there were 50,000 formulations of 1,500 basic chemicals. Khilchevskaya (1980) assessed the former at 100,000 and the latter at 900, claiming that 120 pesticides, out of 240 studied, were genotoxic. Considerable evidence has accumulated on the genetic effects of pesticides on various organisms (Fleck and Hollaender, 1982). Approximately 50% of all studies were on plant systems.

Direct Evidence

In studying components of agroecosystems, ploidy variations, mutations and cytogenetic abnormalities have been shown to occur in a variety of grain crops (Liang et al., 1967; Behera et al., 1982). Fertility losses in sorghum have also been recorded (Liang et al., 1967). Chlorophyll deficient mutations have been reported in barley used as an experimental system (Panda and Sharma, 1979). Resistance to herbicides has occurred in both

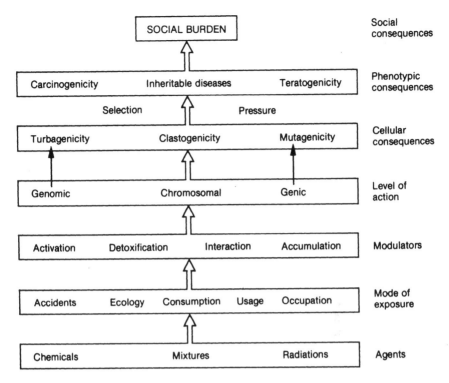

Figure 12.1   Outline of Genotoxicology

crop plants and weeds (Grignac, 1978). Weeds with genotoxic manifesta-
tions have also been recorded from fallow lands with a history of herbicide
application (Tomkins and Grant, 1976).

Genetic changes in rodents in Vietnam have been attributed to herbicide sprays (Orians and Pfeiffer, 1971). United Nations Environment Program (UNEP) has recorded seven species of rodent resistant to rodenticides,
including *Rattus rattus* and *R. norvegicus*. The Food and Agriculture
Organization (FAO) estimated 182 strains of insect and mite to be resistant
to pesticides in 1965, increasing to 364 by 1979. Approximately 223 agricultural pests have developed resistance to nine major groups of pesticides.
Cotton leaf worm developed resistance to ten insecticides during 1961-
1976, in Egypt. The World Health Organization (WHO) has estimated that
the number of resistant mosquitoes and other arthropods increased from
102 to 121 during 1968-1976. Resistant plant pathogens were negligible in
1965, but increased to 35 species after the introduction of systemic fungicides and the acceleration of herbicide use. The genetic bases of this resistance are well established (UNEP, 1979).

A high incidence of mutations among organisms lower on the evolutionary scale is indicated by a review of pesticide effects on microbes, although literature specifically related to mutations is very meagre (Simon-Sylvestre and Fournier, 1979).

In India, following the green revolution, several pests increased in importance. A minor disease, Karnal Bunt, became a major one on wheat; yellow and brown rust assumed epidemic proportions in the high yielding varieties of wheat; gall midges, tungro virus, bacterial blight and brown plant hopper became major pests of rice. Possible explanations are that pests are evolving new strains or that the crop varieties are losing their resistance, under the influence of chemical applications, both factors being under genetic control and therefore inherited.

Indirect and Circumstantial Evidence

The possibility that microbes and mammals respond to pesticides in the same way as do plants is supported by concordance estimation from the published literature. Nearly 80% concordance has been found between plants on the one hand and microbes and mammals on the other (Sharma, 1982). Allowing for differences in rates of exposure, this implies that biota in the agroecosystem may be under genetic stress from pesticides to the same extent as experimental plant systems.

Natural vegetation in habitats receiving industrial effluents showed genotoxicity (Klekowski and Levin, 1979). Similar results were obtained when tester strains were used as monitors (Plewa and Gentile, 1982). Genetic abnormalities have also been found in pesticide workers (Yoder et al., 1973). In Handigodu, Karnataka State, India, about 600 people have been affected by a variety of bone degenerations during the last eight years. The effects are found even among infants and include dwarfism. Consumption of crabs and fish from paddy fields receiving insecticides is epidemiologically linked to these disorders. Similar events on a lesser scale have occurred in other parts of Karnataka State.

CONCLUSIONS

Slow and imperceptible changes may occur in the biota of an agroecosystem under pesticide stress. Pesticide inputs are low in those tropical ecosystems which are most rich in biota, but their use is increasing. Indications of the breakdown of agro-ecosystems may be: crops are becoming more susceptible to pests; pests are developing resistance to pesticides. Chemical inter-actions between pesticides and fertilizers may lead to new and more toxic compounds. Natural fertility of the soil due to microbes may be depleted. The breeding system of the crop may be modified. Habits and habitats of useful organisms may be altered, affecting host-parasite predator-prey relationships.

Various factors could either accelerate or retard the development of genetic stress from pesticides, and knowledge of these moderating factors is fragmentary. Overdosing with pesticides is a potential accelerating influence, while pesticide adulteration diminishes effectiveness against the target species, leading to the evolution of new strains. Many pesticides contain noxious impurities (Grant et al., 1976). If genotoxic pesticides are residual in nature they pose a long term hazard in food chains, to an extent dependent on the chemical's half life. Plants may also activate nongenotoxic pesticides into mutagens (Plewa and Gentile, 1982). Degradation generally removes the hazard, but can generate more dangerous compounds as in the case of diflubenzuron and 1,2-dibromoethane (Scott et al., 1978). Dilutents, adhesives, adjuvants and wetting agents are generally considered genetically inactive.

Developing a proper perspective on imperceptible mutation pollution by pesticides in agroecosystems, with ramifications through the food chain, is a necessary part of environmental monitoring for pesticide hazards.

REFERENCES

Behera, B.N., Sahu, R.K. and Sharma, C.B.S.R. (1982). Cytogenetic hazards from agricultural chemicals: 4. Sequential screening in the barley progeny test for cytogenetic activity of some systemic fungicides and a metabolite. *Toxicology Letters* 18: 195-203.

Crow, J.F. (1983). *Identifying and Estimating the Genetic Impact of Chemical Mutagens*. Washington, D.C.: National Academy of Sciences.

Fleck, R.A. and Hollaender, A. (ed) (1982). *Genetic Toxicology: an Agricultural Perspective*. Plenum. London, New York.

Grant, E.L., Mitchell, R.H., West, P.R. and Ashwood, S.M.D. (1976). Mutagenicity and putative carcinogenicity tests of several polycyclic aromatic compounds associated with impurities of the insecticide methoxychlor. *Mutation Research* 40: 225-8.

Grignac, P. (1978). The evolution of resistance to herbicides in weedy species. *Agroecosystems* 4: 377-85.

Khilchevskaya, R.I. (1980). Environmental pollution and genetic information. *Impact* 30: 191-6.

Klekowski, E.D. and Levin, D.E. (1979). Mutagens in a river heavily polluted with paper recycling wastes: results of field and laboratory mutagen assays. *Environmental Mutagenesis* 1: 209-19.

Kurinnyi, A.I. and Pilinskaya, M.A. (1974). Pesticides as a mutagenic factor in environment. *Tsitologia et Genetica* 8: 342-73.

Liang, G.H., Feltner, K.C., Liang, Y.T.S. and Morril, J.M. (1967). Cytogenetic effects and responses of agronomic characters on grain sorghum following atrazine application. *Crop Science* 7: 245-8.

Orians, G.H. and Pfeiffer, E.W. (1971). Ecological effects of the war in Vietnam. *Science* 168: 544-54.

Panda, B.B. and Sharma, C.B.S.R. (1979). Organophosphate induced chlorophyll mutations in *Hordeum vulgare*. *Theoretical and Applied Genetics* 55: 282-284.

Plewa, M.J. and Gentile, J.M. (1982). The activation of chemicals into mutagens by green plants. In *Chemical Mutagens* (eds F.J. de Serres and A. Hollaender) 1: 401-420.

Ridgeway, R.L., Tinney, J.C., Mac Gregor, J.T. and Starler, N.J. (1978). Pesticide use in agriculture. *Environmental Health Perspectives* 27: 103-12.

Scott, B.R., Sparrow, A.H., Schwemmer, S.S. and Schairer, L.A. (1978). Plant metabolic activation of 1,2-dibromoethane (EDB) to a mutagen of greater potency. *Mutation Research* 49: 203-212.

Sharma, C.B.S.R. (1982). Plant monitors of environmental mutagens. In *Workshop on Evaluation of Mutagenic and Carcinogenic Potential of Environmental Agents*, 27-39. Bombay: Environmental Mutagen Society of India, BARC.

Simon-Sylvestre, G. and Fournier, J.C. (1979). Effects of pesticides on the soil microflora. *Advances in Agronomy* 31: 1-92.

Tomkins, D.J. and Grant, W.F. (1976). Monitoring natural vegetation for herbicide induced chromosomal aberrations. *Mutation Research* 36: 73-84.

United Nations Environment Programme (1979). *Annual Report*, 7-10.

Yoder, J., Watson, M. and Benson, W.W. (1973). Lymphocyte chromosome analysis of agricultural workers during extensive occupational exposure to pesticides. *Mutation Research* 21: 335.

PART II
PEST MANAGEMENT DECISION MAKING
AT THE COMMUNITY LEVEL

13

Philippine Rice Farmers And Insecticides: Thirty Years Of Growing Dependency And New Options For Change

P.E. Kenmore, J.A. Litsinger, J.P. Bandong, A.C. Santiago and M.M. Salac

INTRODUCTION

For the last thirty years, Philippine rice farmers have used more and more insecticides, despite receiving neither higher yields nor profits. Since studies began in farmers' fields in 1976, yields and profits have not significantly increased yet over 90% of farmers now use insecticides on every crop; they perceive them to be essential and act accordingly. This paper discusses how Philippine rice farmers came to depend on insecticides, whether this dependency is needed to keep up their rice production or income, and how since 1974, they have begun training for self-reliance in a practical alternative, integrated pest management (IPM).

GROWTH OF INSECTICIDE DEPENDENCY

Changes in insecticide use among Philippine irrigated rice farmers from 1954-1984 are shown in Figure 13.1. Irrigation makes two rice crops a year possible. Whilst only about one third of Philippine rice land is well irrigated, this portion produces over two thirds of the rice in any given year. Insecticide use in the early 1950's was very low, but the next decade saw a sharp rise so that in 1965 *before* the release of the first green revolution variety, IR8, about 60% of irrigated rice farmers were using insecticides.

In the decade after 1965, the rate of increase slowed down so that by 1976, just under 90% used insecticides. By the mid 1980's, over 95% of irrigated rice farmers used insecticides on every rice crop they grew. These data suggest that factors other than either the release of green revolution rice varieties or the rice intensification campaign (Masagana 99), begun in 1973, were responsible for convincing the bulk of irrigated rice farmers to use insecticides.

Part of the explanation is given in Figure 13.2, showing the number of hectares of irrigated rice treated free of charge by technicians of the Bureau

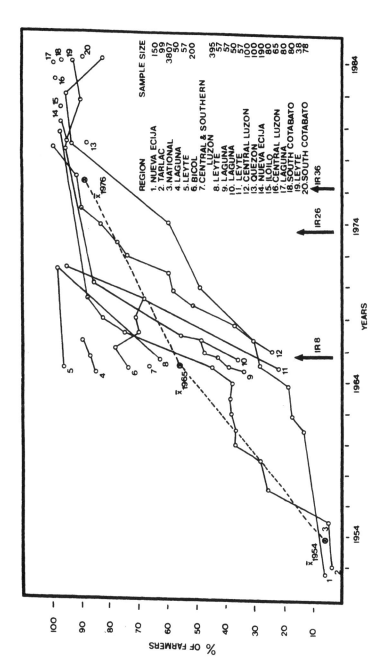

Figure 13.1 Proportions of Philippine rice farmers, in villages with irrigation, using insecticides 1954–1984

Key: ——— Changes among the same rice farmers in the areas specified (sample size), – – – Mean proportions for 1954, 1965 and 1976, calculated as the unweighted average of village scores; ◄ The first release of green revolution rice varieties of Plant Industry (BPI). This increased to over 270,000 ha in 1965, over half the irrigated rice land at that time. Chemicals were purchased by the government or included as aid in bilateral programs. Some were given to the government by chemical companies interested in helping Philippine farmers modernize, and in developing their markets for pesticides. There was a general atmosphere that better farming meant using more inputs: "Filipino farmers are just now beginning to accept scientific methods. Their inherent resistance to change is gradually breaking down. The use of fertilizer and other agricultural chemicals is bound to multiply. Once the wheel of progress moves, it is likely to gain more momentum." (Von Oppenfeld, 1958). In contrast to the government's interest in providing free chemical control, the BPI program of releasing *Trichogramma* parasites to control the major pest of that time, rice stemborers, declined and finally ended altogether. Insecticides were effective, easy to store, easy to deploy in the field, and were a more negotiable commodity for companies, rural retailers, rural patrons, farmers, and technicians. It is surprising, in comparing Figures 13.1 and 13.2, that farmers lagged by nearly a decade in adopting insecticides. Technicians recall that farmers were not willing to use insecticides until repeated demonstrations were made, suggesting that it would now take another decade of aggressive extension to shift farmers to more effective pest control technology such as IPM.

Over the 30 years concerned farmers also spent more and more money on insecticides. The proportion of the total cost of irrigated rice farming spent on insecticide increased from below 1% to over 10%. A study by the Agricultural Economics Group of the University of the Philippines, Los Banos, on 'The Cost of Producing Palay (Rice) in Laguna' followed one group of farmers for eight cropping seasons and found an explosive increase in their insecticide investment, from 3.5% to 15.5% of the cash cost of production in four years. At the end of this period, farmers reported a widespread outbreak of rice brown planthopper, *Nilaparvata lugens*, that wiped out the total production of about 20% of the farmers included in the study. This came after their increase in insecticide use, as would be expected from field studies of the regulation of brown planthopper populations by natural enemies (Kenmore et al., 1984). Laguna farmers sprayed so much partly because credit and chemicals were more available but principally because, during the first year, irrigation allowed double cropping in much of Laguna for the first time. While the dramatic outbreak of brown planthoppers did not occur until insecticide use intensified, pests like stemborers that respond directly to cropping intensity did increase, presumably inducing farmers to spray (Loevinsohn, 1985). As the brown planthopper became a nationwide pest, the government responded by increasing the area it treated (Figure 13.2), but this was not effective and the pest was eventually controlled by the release of the resistant variety IR36, which kept brown planthopper populations down in spite of farmers' continuing to use more insecticides.

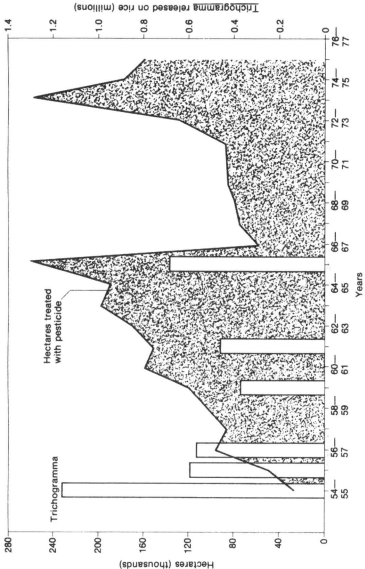

Figure 13.2   Responses to insect pests on rice

Key: ———— Thousands of irrigated rice hectares treated with insecticides by staff of BPI between 1955 and 1977, when treatments stopped;   ☐Numbers of *Trichogramma* parasitic wasps released by the BPI to control rice pests, especially rice stemborer, between 1954 and 1966, when the project stopped

ATTITUDES TOWARDS INSECTICIDES

Surveys of farmers' attitudes show that since 1980 they have accepted insecticides as an irreplaceable part of modern rice production (Canedo, 1980; Corado, 1985; Escalada, 1985; Espina, 1983a and Oliva, 1984). They believe that treating crops with insecticides is progressive, modern, effective, and necessary; that farmers who use insecticides are themselves progressive, modern, and hardworking; that farmers who do not use insecticide are lazy, old-fashioned, and ignorant; that insect pests can cause major yield losses, cause those losses before the insects are visible, and can be controlled by chemicals; and that using more insecticide ensures higher yields and profits even though insecticides are dangerous. On the whole, more farmers use insecticides than use fertilizers (Canedo, 1980; Corado, 1985; J.A. Litsinger, unpublished data). Espina (1983b) concluded that farmers living in a hazardous environment, subject to floods, typhoons, and occasional drought, saw insect pests and diseases as one part of the environment they could control for themselves, with powerful chemicals, so that this sense of control outweighed the perceived dangers of the insecticides themselves. Farmers believe calendar-based treatments are preferable, but cannot afford them and so treat instead at the first sight of an insect; they believe modern, usually insect resistant, varieties require more treatment than previous varieties; that if their neighbor treats they should, regardless of pest population; that their whole farm should be treated even if pests are confined to one paddy or less than 10% of their area; that they should treat their fields after fertilizer applications; and that insecticides are always more effective than natural enemies (Binamira, 1985; Escalada, 1985; Rivera, 1985).

Notwithstanding this tidal wave of approval for insecticides, some farmers stood firm. Particularly in the far north Ilocos region, farmers retained a number of cultural practices, used native plants as pest repellents or pesticides, and adopted only those new practices that paid off (Rosario-Cardenas, 1981; Carino, 1985; Corado, 1985). Farmers who combined old practices with new had higher returns on investment in pest control than wholehearted pesticide adopters. Often these practices are very specific to the ecology of a particular insect pest. A survey conducted in 1984 of 93 farmers who had practised botanical pest control from five regions of the country found over 70% of them using one plant (*Glyricidium* sp.) in the same manner (broadcasting leaves) at the same crop growth stage (one week after transplanting) to control rice caseworm. This pest cuts a small tube of leaf and matures inside it while floating on the paddy water. Surrounded by a solution of either the leachate of broadcast leaves or synthetic insecticide, it is vulnerable to low concentrations of toxic chemicals. There is no broad spectrum botanical insecticide in the Philippine farmers' ethnoscience and this makes broad spectrum chemical insecticides relatively attractive. All but three of these farmers surveyed had abandoned botanicals for more potent synthetics. However two had recently gone back to botanicals as the price of synthetics had tripled.

ARE INSECTICIDES NEEDED IN EVERY CROP?

Between 1976 and 1984, over 330 farmers' crops were used to compare treated with untreated fields. In only 50% of these was there a measurable yield loss to insects. In irrigated areas growing resistant varieties, starting with IR36, the proportion of fields showing yield loss to insects dropped to 42% (Litsinger, 1984). From 1980-1983, an additional 105 farmers' crops were followed in double cropped irrigated areas; again only 50% showed a significant yield loss (Sumangil, 1984). In all cases the predominant varieties planted by local farmers were studied. This means that while about 97% of farmers use insecticides, only 50% get higher yields from using them.

Part of the problnem is misidentification. Marciano et al. (1981) found that among farmers in Laguna, the most modern rice growing province, insecticide use was not correlated with insect damage, but was significantly correlated with *disease* severity, including fungal and bacterial diseases which are not affected by insecticides.

This picture of Philippine rice farmers' problems with perception and management of insect and disease pests does not indicate that they cannot manage other pests. All surveys cited showed that farmers managed weeds as effectively and more profitably than researchers. They understood the importance of timing weed control early in the season, and how to combine cultural, varietal, mechanical and chemical weed control methods in an elegant IPM system. They used herbicides judiciously and profitably.

Farmers also manage rat control well. They hold mass rat drives, destroy burrows, try to synchronize their harvesting schedule and use acute and occasionally chronic rodenticides, if these are available, in effective ways adjusted to local village conditions. It is an interesting perceptual point that farmers do better at controlling large visible pests than small, or microscopic ones (insects and diseases) whose damage symptoms are often the only indication that something is wrong.

Perhaps the problem is economic. Farmers believe increased insecticide use means increased profits. Need-based use of insecticides was compared with calendar-based use by Litsinger (1984) and Sumangil (1984), in the same fields with yield loss trials. In 80% and 77% of these trials respectively, the use of thresholds was more profitable than calendar treatments. Herdt et al. (1984) used historical data from regional experimental stations and farmers' fields to show that while minimum use of insecticides was more profitable than no use, *any level above minimum* (at least two levels above minimum were used in each trial) *reduced* the chance of profit, compared to minimum. Beliefs that insecticides are necessary for high yield and that a higher frequency of treatment is more profitable than a lower frequency are not supported by field results.

Farmers cannot afford to use higher frequencies and dosages of insecticides, even though they believe it would benefit them. In 1984, 31 village sites, each containing three farmers' field replicates, were maintained in the wet or main growing season. We compared yields, total costs, costs of pest control and percentages of total costs spent on insect control for farmers' practice and for IPM using action thresholds. Table 13.1 shows that IPM preserved the same yield, while reducing total costs of pest control by more than 50%, or reducing cash costs (excluding inputed labor cost) by 80%.

TABLE 13.1

Costs and yields of different insect control treatments. Philippines national survey, wet season 1984

Treatment	Cost of Insect Control (P/ha) † Mean		Total Costs (P/ha) S.D.		Cost of Insect Control (% total)	Average Yield (Tons/ ha)
IPM	158	54	3611	1625	4.4%	5.07
Farmers' practice	351	185	3905	1721	9.0%	4.90

Note: Data from 43 farmers in five regions. Costs of IPM treatments include P100/ha as season-long imputed opportunity cost of monitoring so actual cash costs on the order of P58/ha.

† P = Pesos

EXTENDING THE IPM ALTERNATIVE

Clearly, alternative insect control technology, using insecticides on a need basis with proper identification, could save rice farmers money and improve the benefits derived from insect control. Also the annual costs of rice insecticides in the Philippines is about $8 million, in foreign exchange. To encourage less dependence and greater self reliance, the Philippine National IPM Program, with FAO support, has developed a method of extension that uses field demonstrations and farmers' training and includes ten steps:

i)   site selection concentrating on high insecticide use areas as pilot sites;

ii) village entry through local officials or elders;

iii) mass meeting or scaring session to provoke interest in farmers' classes;

iv) baseline survey of 25 farmers' practices, costs, and returns; this is summarized and posted in the village hall;

v)   a)   selection of three farmers to hold result demonstrations in their fields;

b)	farmers' class meetings — total of 40 hours;
vi) a)	set up result demonstrations;
b)	field visits 'on call' to farmers reporting pests;
vii) a)	weekly monitoring of result demonstrations;
b)	weekly field method and demonstrations emphasizing proper identification and need-based use of insecticides;
viii) a)	evaluation of farmers' skills in the field;
b)	evaluation of farmers' conceptual knowledge in classrooms;
ix) a)	yield and economic analyses of results and demonstrations;
b)	final discussions and planning future farmer IPC group activities; arranging for follow-up visits;
x) a)	publication (via posters) of results from field;
b)	graduation ceremonies with local officials and elders.

Since 1979, over 3000 farmers in 117 villages in all 12 regions of the Philippines have been trained; in 1985, 4981 extension workers had 20 hours of training in how to set up village IPM sites. National, regional, provincial and municipal governments have all contributed to the funding of these efforts.

EVALUATING IPM EXTENSION

Three years after two such sites were established and the farmers' classes completed, we conducted a survey comparing trained and untrained farmers from the two villages. There were significant differences between trained and untrained farmers' knowledge of:

● sources of insect infestation;

● names of major pests of the current and immediately preceding season;

● descriptions of these major pests;

● how to use action levels in decision making;

● names of important natural enemies;

● effects of insecticides on natural enemies;

● safety precautions for insecticide application.

There were significant differences in the understanding of the two groups of farmers regarding:

● insecticide requirements of resistant varieties (trained farmers said less required);

● the need to treat when a neighbor treated (trained farmers said not);

● the need to treat after fertilizing (trained farmers said not);

- 105 -

- the profitability of need-based use (trained farmers said it was more profitable).

Trained farmers also reported significantly fewer insecticide applications per season than untrained farmers (3.5 compared to 5.7).

While knowledge and understanding are important indicators of eventual behavior, interview surveys cannot illustrate farmers' grasp of field skills. In 1984 we developed a method of comparing the performance of trained and untrained people in pest identification, recognition of action levels, choice of chemicals, natural enemy identification, and limited coverage insecticide use (applying chemicals only to those field spots actually infested). It is a modification of a laboratory practical exam, administered in the field, with pest specimens and damaged plants in roughly their naturally occurring positions. Answers are recorded by voting, to minimize the use of abstractions and reduce literacy requirements. (We are planning a trial with audio cassettes to replace printed questions entirely.) As indicated in Table 13.2, the scores for trained and untrained farmers do not even overlap in range, suggesting the kind of skills being taught are novel.

TABLE 13.2

Scores of trained and untrained farmers from the municipality of Candaba, Pampanga on a ballot-box field test of IPM skills

Score Ranges	Trained Farmers	Untrained Farmers
13 — 15	1	0
10 — 12	13	0
7 — 9	4	0
4 — 6	0	17
1 — 3	0	2
Mean	10.5 (70%)	4.7 (31%)

One possible source of this novelty is the concentration on individual plants as the unit of identification for diagnosis and decision making. In a recent study of 60 farmers' decision making (Bandong et al., 1985), data were gathered during field walks with each farmer every two weeks during the season, and a dialogue on the state of the field and that week's pest control decisions was held. If a decision to treat had been made, the actions of observation were recounted and acted out by the farmer. The farmers made many decisions based on the general appearance of field plots, both their neighbors' and their own. They classified fields by elevation, low-lying fields holding more water, more vegetation, and more insects. They examined field edges before field centers, and used categories for the aspect of each whole paddy to describe damage levels. Only after 3-6 prior decision points were passed, did about 15% of the farmers' decisions include the condition of infestation levels of *individual* plants (these farmers had intensive IPM village training). In these decisions individual plants were examined and a rough summation of damage performed, and it is possible that, in addition to the novelty of counting, the *perceptual* leap to examining individual plants and extrapolating from a sample to decisions about the whole field was even more of a novelty. This also may explain farmers' reluctance to treat only infested spots in their fields — perhaps the perceptual unit is the whole parcel, and if one spot is infected, the whole parcel is presumed contaminated. We are planning to address the problem of learning to focus on single plants and extrapolate to the field in a set of field exercises.

The ballot box field test was also used to evaluate extension technicians. During the national training held in June–August 1985, a subsample of 1142 technicians was given a standard 'ballot box' test *before* training. The average score was 62% correct, far below the level expected of those who will train farmers. The areas of greatest weakness were natural enemy identification, disease identification, use of action levels and spot treatment. Follow up training will concentrate on these areas. After one, two, four, and perhaps more crop seasons we plan to retest the technicians to estimate retention and reward those with the highest skills levels.

The methods described here have produced significant improvements in trained farmers' knowledge and attitudes and new field skills evaluation tests have been successfully used with technicians and farmers to point out weaknesses in training and estimate the impact of training. Armed with these extension tactics the Philippine National IPM Program now has a strategy to match the effort of 30 years ago, this time extending a pest control technology that is more profitable, stable, safe, and self-reliant.

REFERENCES

Bandong, J.P., Litsinger, J.A. and Kenmore, P.E. (1985). Farmers' management of rice pests: Its implications to IPM implementation. Paper presented at 15th Annual Meeting, Pest Control Council of the Philippines, May 1985.

Binamira, J.S. (1985). Marketing study for the integrated pest control technology. Paper presented at the FAO IPC Impact Meeting, August 1985.

Canedo, F.M. (1980). Change agents' perceived credibility and their influence in the innovation-decision process of development program. Ph.D. dissertation on Community Development, U.P. at Los Banos, Philippines.

Carino, F.O. (1985). The indigenous and recommended pest control practices of rice pests: The case of the Zanjera farmers' organizations in Espiritu, Ilocos Norte. Paper presented at the FAO IPC Impact Meeting.

Corado, M.E. (1985). Village level analysis and technology dissemination: A case in Leyte. M.S. thesis, U.P. at Los Banos, Philippines.

Escalada, Monina (1985). Rice farmers' knowledge, attitudes and practice of integrated pest control (Western Leyte). Final report submitted to the FAO Intercountry Program on Integrated Pest Control (Rice).

Espina, M.L.P. (1983a). Some methodological guides in conducting interviews with women: A brief glimpse on the participation of women technicians in extension work. Paper presented at NCPC/FAO workshop on the role and potential of the Filipina in rice crop protection, SEARCA.

Espina, M.L.P. (1983b). Rice farmers' perceptions and responses to pest hazards. M.A. thesis, Ateneo de Manila University, Philippines.

Herdt, R., Castillo, L. and Jayasuria, S. (1984). The economics of insect control in the Philippines. *Judicious and efficient use of insecticides*, IRRI, Los Banos, Philippines.

Kenmore, P.E., F.O. Carino, Penez, C.A., Dyck, V.A. and Gutierrez, A.P. (1984). Population regulation of the rice brown planthopper (*Nilaparvata* Stal) within rice fields in the Philippines. *Journal of Plant Protection in the Tropics* 1(1): 19-37.

Litsinger, J.A. (1984). Assessment of need-based insecticide application for rice. Paper presented at MA-IRRI Technology Transfer Workshop.

Loevinsohn, M. (1985). Agricultural intensification and rice pest ecology: Lessons and implications. Paper presented at the International Rice Research Conference.

Marciano, V.P., Mandac, A.M., and Flinn, J.L. (1981). Insect management and practices of rice farmers in Laguna. Paper presented at Crop Science Society of the Philippines Annual Meeting.

Oliva, A. and Imperial, S. (1984). Farmers' pest control attitudes and practices in two Bikol villages. A report of the Research and Service Center, Ateneo de Naga University.

Rivera, M.T.M. (1985). A survey on rice farmers' knowledge, attitude, and practice (KAP) on crop protection in Northern Mindanao. Paper presented at the FAO IPC Impact Meeting.

Rosario-Cardenas, V. (1981). The coopetation of indigenous and recommended rice technologies among the Ilocanos. Ph.D. dissertation, Extension Education, U.P. at Los Banos, Philippines.

Sumangil, J.P. (1984). The integrated pest control component of cost reducing technology for Masagana 99 in the Philippines. Paper presented at the Agritech 84 Conference, Manila, Philippines.

Von Oppenfeld, H. (1958). Nigrogen consumption: An index of agricultural development in Asian countries. *Philippine Agriculturist* 42: 278-280.

14
The Motivating Factors For Community Participation In Vector Control

Jean Mouchet and Pierre Guillet

INTRODUCTION

Man has always been aware of the nuisance of blood sucking arthropods and ectoparasites, and delousing was a common practice in most ancient cultural systems. In 2000 B.C. the Chinese already used pyrethrum powder against ectoparasites. Sulphur treatments for scabies have been known for several centuries in European pharmacopoeia. However, only at the end of the 19th century was the role of insects in the transmission of diseases discovered, following the advances of modern science and the discovery of pathogenic infectious agents. This knowledge has spread from scientific and medical circles to the general population according to the degree of impregnation by western scientific culture.

Whereas parasite control was the responsibility of the individual vector control has been organized by specialized bodies at provincial, national or international levels. General sanitation measures against vectors of malaria, filariasis and yellow fever were implemented before the organization of large vertically integrated structures like the Malaria Eradication Program. To begin with in tropical countries, only individuals in the community were involved in measures to reduce the sources of disease, e.g., destruction of domestic and peridomestic mosquito breeding sites, in response to legislation dictated under colonial rule. The vertical programs were carried out with very limited participation by the populations benefitting from them, to the extent of welcoming spray teams into their homes and sometimes providing water. When the population was not willing to collaborate this led to very serious impedments to spraying operations.

The development of new health policies, based on primary health care, implies that the population can no longer be a passive bystander of health service teams, but should become an active participant in vector control (Mouchet, 1982).

CONDITIONS AND LIMITS OF COMMUNITY PARTICIPATION IN VECTOR CONTROL

Community Participation and Health Service Activities

Vector control operations exhibit varying degrees of complexity. For example measures like weekly cleaning of drinking water containers for *Aedes aegypti* control require limited technical skills, whereas weekly treatment of rivers for onchocerciasis vector control requires the use of helicopters over very large areas ($800,000$ km^2). The first type of activity can be left entirely in the hands of the community, but the second must be organized and carried out within the large and specialized framework of a vertical program.

Community participation should be complementary to the action of health services. The role of each partner depends on the complexity of the strategies and techniques to be applied, and on the operational capabilities of the community. The latter can be improved by personnel training, increased financial resources and measures to promote willing participation.

Control methods, including individual protection by nets and mosquito coils, which can be transferred to communities alone or integrated into a program, were reviewed by the 7th WHO Expert Committee on Vector Biology and Control (Anon, 1983). Vaughan (1980) has also reviewed relationships between communities and health services and the status of community agents.

This paper discusses the requirements for community adhesion to vector control activities and for persistence of motivation.

Community Motivation

All community participation relies on knowledge of the role of the vector and also of the disease which it transmits, in order to generate understanding within the population of the objectives of the proposed activities. Health education, which should take into account the cultural background of the population, is the essential tool for dissemination of this knowledge. Any proposed action should be compatible with local cultural habits and local community representatives should be consulted.

Tedious repetition of the same activities is often necessary for long term prophylactic measures, and this creates a high risk of discouraging the population. It is difficult to maintain the community interest for long term activities, and continuity in such efforts is more likely to be achieved if the vector is also a nuisance in its own right, or if its control also kills other pests. Limiting the nuisance improves welfare and creates a favorable impression of the results of the actions undertaken.

In the past, several potentially efficient malaria control campaigns have been impeded by refusals to allow access to houses because the insecticide (DDT) was no longer killing DDT-resistant bedbugs or *Culex*, and the treatment was considered as useless. In Sao Tome, it was necessary to add malathion to DDT to kill the bedbugs, in order to make the campaign acceptable (Viegas de Ceita, pers. comm.). If such an attitude arose in a population whose participation in a program was essential, more serious problems would arise.

Ways of Participation

The community must be supported by health services and by other government agencies, to make their public health actions compatible with other activities. For example it would be useless to remove empty containers acting as *A. aegypti* breeding sites without proper disposal by the municipalities. The breeding sites would just be displaced from the house to the nearby refuse dump (Yebakima et al., 1979).

Some reports have stressed the need for adequate legislation (Anon, 1983), but although such legislation exists in many countries it is often not enforced. It relies on constraints which have an adverse effect on the spirit of participation. A determined effort can be made by benevolent individuals for a short time in emergencies, but long term prophylactic activities rely mainly on permanent or temporary health agents paid by the community or other bodies. Complementary work requested from ordinary members of the community should be as light as possible. Ideally, simple sanitation measures would become an integral part of day to day behaviour.

Given the constraints on third world communities, it is an ethical duty for planners to propose only activities that are highly effective in relation to the efforts required. Control measures proposed to a community must be defined according to the local epidemiological characteristics of the disease and the local ecology of the vectors and very few control methods can be generalized. Elaboration of locally adapted strategies and techniques to develop a comprehensive program which is more than mere enumeration of routine classical measures, the efficacy of which could be questionable in a given situation, requires a high degree of local expertise. Such expertise is not yet available in most countries and good training and a career network are necessary for its development (Mouchet, 1982; Anon, 1983).

REPRESENTATION OF THE ROLE OF VECTORS AND OF THE ENVIRONMENT

The representation of disease was part of the philosophical systems of societies, elaborated long before the discoveries of modern western science. In many African cultures, such as the Congo, disease was not an entity but a particular condition of man often due to a dissention with his environment or to magical practice. The role of vectors was unknown.

The present representation of disease, and of the role of vectors, is a compromise between traditional concepts and western culture. It varies from one social class to another, according to the level of education, the level of development of non-traditional medical structures, and to disease exposure.

On the Niari district of the Congo, between 1910 and 1940, sleeping sickness became very prevalent and killed a large part of the population. Its efficient control was due to strong medical action, the basis of which is still functioning. The disease is called *nberre tolo* and the role of tsetse flies is well known (they are even inserted in cartridges shot on the graves of patients who have died from sleeping sickness, to avoid the 'evil spirits' considered responsible). On the borders of the endemic focus, where the disease is less severe, the representation of the vector is less clear. All biting insects, even midges, are considered equally responsible. The adoption of traps for tsetse control by villagers of the Niari is largely due to their knowledge of the vector.

In Northern Cameroon in 1962 the Chaos seem to have been aware of the relationship between tsetse flies and a lethal disease of cattle. As soon as tsetse disappeared after insecticide spraying (Mouchet et al., 1961), they brought their herds to the forest gallery along the Logone river, a site which they had previously avoided. It was not possible to establish whether this was based on recent knowledge or a traditional belief.

In urban Pondicherry, India, a survey showed that 80% of the people linked mosquitos to the clinical symptoms of bancroftian filariasis; 9% blamed supernatural forces; 4% blamed polluted water (Anon, 1981). However, along the coast of East Africa, neither Arabs nor Bantous linked filariasis and mosquitoes, although the nuisance of *Culex quinquefasciatus* led to requests for control measures (Subra, pers. Comm.).

In Guinea some people consider that onchocercian blindness is due to the blood of engorged blackflies squashed on the face when biting. This may be a recent interpretation resulting from a misunderstanding of the scientific explanation.

Malaria is well known in most of tropical Africa as 'warm body'. In the Congo and in Burkina Faso 30-50% of febrile cases in children are due to malaria (Collective, 1982; Carnevale et al., 1983; Richard et al., 1984). In most ethnic groups, the role of mosquitos is well known, and sometimes the term *Anopheles* is mentioned, but people are generally unable to differentiate Anophelines from *Culex*, *Aedes* or *Mansonia*. Consequently they consider the persistence of these other mosquitos as a failure of house spraying treatments for malaria control, even if Anophelines have been efficiently controlled.

Presently there is a lack of information on the representation of the role of vectors in such populations and when it exists it is dispersed and anecdotal, which makes its compilation difficult.

Representation of the Role of the Environment

In the literature it has long been known that some areas were regarded by populations as unhealthy, and subsequently some such areas were shown to be sources of disease.

The Greek physician Hippocrates, during the 4th century B.C., recommended establishing villages away from marshes to avoid fevers such as malaria. In certain mountainous areas of Central Africa, where no malaria transmission occurs, spending the night in the surrounding valleys was considered dangerous and was forbidden by tradition. Now it is known that these valleys are highly malarious, and spending a night in them could be fatal for non-immune peoples from the mountains (Mouchet and Gariou, 1960).

The link between onchocercian blindness and rivers was commonly established in Africa, even for those who did not know the natural history of the disease, but the history of this concept that "river eat the eyes" is not established. In savannah areas of West Africa people left their villages close to rivers when symptoms of the disease became too severe and sometimes returned when they began to forget (Hunter, 1966; Lefait, 1976). Burkina Faso populations still have vivid memories of sleeping sickness epidemics in the first half of this century and they consider some tsetse fly infested valleys as evil, in spite of the absence of new disease cases for more than 20 years.

The relationships between the disease and the environment is an important factor to be taken into account when vector control operations are implemented for the recolonisation of deserted or empty lands.

VECTOR CONTROL AND DEVELOPMENT

Development rapidly changes the environment and population distribution of a country, leading to the introduction of new pathogenic agents in non-infected areas and to the arrival of non-immune populations in endemic areas. Very often development infrastructure like dams and irrigations systems, lead to an increase in vectors and pests (Philippon and Mouchet, 1976). Structures built as part of new development schemes must take account of vector and general public health problems. For example the design of dam spillways and irrigation channels can avoid or limit blackfly and snail populations. Now, in many communities, people with expertise gained in agriculture, and suitable equipment, can be found to participate in public health activities after only a short training. In Indonesia, a trial was carried out by temporary workers for the control of *Anopheles aconitus* in cattle shelters (Barodgi et al., 1984) and it seems to have been successful. In Mali, on the Bandiagara Plateau, community first aid agents participate in the surveillance of onchocerciasis for the Onchocerciasis Control Program (Anon, 1984).

An increase in vector density does not always imply an aggravation of the disease load. For example in rice field villages near Bobo-Dioualsso in

Burkina Faso the number of anopheline bites per person per night (mainly *A. gambiae* s.s.) is five times more than in nearby villages practicing pluvial culture. However, as the sporozoite rate in the rice field area is several times less than in the other villages, malaria transmission and infant and child mortality were lower (Carnevale et al., 1983). This may be partly due to a higher standard of living. This underlines the necessity for detailed epidemiological studies before planning and implementing vector controls.

Galloping urbanization, a major trend in human ecology, is an uncontrolled phenomenon. Sanitation cannot keep up with the exponential growth of cities, leading to the proliferation of some anthropic vector species such as *C. quinquefasciatus* and *A. aegypti*, and to a declining quality of habitation. A large scale trial has been undertaken in Pondicherry, to evaluate integrated control measures against filariasis vectors, applied with municipality and community participation (Rajagopalan and Paniker, 1984).

The immigrants in large towns have left their traditional social structures and very often have not been integrated into a new framework. Requiring their participation in any health-related activity has to deal with this social disorganization.

CONCLUSION

Community participation is not merely a cheap panacea for vector control problems which cannot be solved by health services. The whole vector control system should be coherent and community and health service activities should be as complementary as possible. The role of each depends on the tasks to be accomplished and the capacity of the community to realize them. Even in national or international vertically integrated programs, maximum effort must be made to motivate the community. Its participation should gradually increase, the final goal being to put in its hands the appropriate tools and expertise to solve most of its vector control problems and to control vector borne disease.

To reach such a goal requires adequate training at all levels. On technical grounds, control methods should be as simple and efficient as possible, and the work of community agents should concentrate on essential activities which have already proven their usefulness.

ACKNOWLEDGMENTS

We wish to thank F. Hagenbucher for unpublished information on the Congo, Dr. P. Brey, Institut Pasteur, Paris, and Mrs. M. Teppaz, ORSTOM, who prepared the manuscript.

REFERENCES

Anon (1981). Vector Control Research Center, Pondicherry, *Annual Report*.

Anon (1983). *Integrated vector control*, 7th Report of the World Health Organization Experts Committee on Vector Biology and Control. *Technical Report Series no. 688* WHO Geneva.

Anon (1984). *10 ans de lutte contre l'onchocercose en Afrique de l'Ouest*. Onchocerciasis Control Program, Ouagadougou, Burkina Faso, ed. Issue 84. 3.

Barodgi, Shaw, R.F., Prad'han, G.D., Bang, Y.H. and Fleming, G.A. (1984). Community participation in the residual treatment of cattle shelters with Pirimiphos-Methyl (OMS 1424) to control a zoophilic malaria vector, *Anopheles aconitus*: large scale field trials. Mimeog. doc. WHO/VBC 84,897.

Carnevale, P., Hervy, J.P., Rober, V., Hurpin, C., Baudon, D., Brandicourt, O., Boudin, C., Ovazza, L., Molez, J.F. and Bosseno, M.F. (1983). La transmission du paludisme dans un périmètre rizicole et en zone de savane de Haute Volta. *Comptes Rendus de la 2ième Conférence Internationale sur le Paludisme et les Babésioses, Annecy, 12-22 Sept*. 1983: 32-42.

Collective paper (presented by J. Mouchet) (1982). Le paludisme en zone rurale au Congo. *De la Géographie à l'Epidémiologie Colloque Tropiques et Santé* CEGET Bordeaux: 109-119.

Hunter, M.H. (1966). River Blindness in Nangodi, Northern Ghana. An hypothesis on cyclical advance and retreat. *The Geographical Review*, 56: 398-416.

Lefait, J.F. (1976). Aspects clinique, épidémiologique et psychosocial de l'onchocercose en zone de savane Africaine, dans la région de Bamako. Thèse Doctorat en medécine, Université de Marseille, 1976.

Mouchet, J. (1982). Vector Control at Community Level. Mimeogr. doc. WHO/VBC/82.847.

Mouchet, J. and Gariou, J. (1960). Anophélisme et Paludisme dans le départment Bamiliké. *Recherches et Etudes Camerounaises*, 1: 92-114.

Mouchet, J., Delas, A. and Yuore, P. (1961). La campagne expérimentale de lutte contre *Glossina tachinoides* à Logone Birni (République du Cameroun et République du Tchad). *Bulletine de la Société de Pathologie exotique 54*: 875-892.

Philippon, B. and Mouchet, J. (1976). Répercussion des aménagements hydrauliques à usage agricole sur l'épidémiologie des maladies à vecteurs en Afrique intertropicale. *Cahiers du CENECA. Colloque International Paris 3-5 mars 1976*, Document 3.2-13.

Rajagopalan, P.K. and Paniker, K.N. (1984) Feasibility of community participation for vector control in villages. *Indian Journal of Medical Research*, 80: 117-124.

Richard, A., Molez, J.F., Carnevale, P., Mouchet, J., Lallemant, M. and Trape, J.F. (1984). Epidemiology and clinics of malaria in villages of the Congo Forest. *Abstracts of the XI th Congresses of Tropical Medicine and Malaria, Calgary, Canada:* 126-127.

Vaugham, J.P. (1980). Barefoot or professional? Community Health Workers/in the Third World. *Journal of Tropical medicine and Hygiene,* 83: 3-10.

Yebakima, A., Schunt, G., Vernerey, M. and Mouchet, J, (1979). Situation d'*Aedes aegypti* en Martinique et considerations sur la stratégie de lutte. *Cahiers ORSTOM, Serie Entomologie médicale et Parasitologie,* 17: 213-221.

15

Agricultural Pests And The Farming System: A Study Of Pest Hazards And Pest Management By Small-Scale Farmers In Kenya

Abraham C. Goldman

INTRODUCTION

This paper is based on a series of farmer surveys conducted in Kenya during 1982 and 1983 to: determine the nature and impacts of pest hazards in peasant farming systems; explore farmer's responses; and examine some aspects of pesticide use on cotton and coffee crops. The underlying theme of the paper is the need to consider pest management within the context of the peasant farming system. Without such a focus, it can be difficult to understand the reasons for many agricultural practices and to design and introduce improvements.

The two study areas were Makueni Location, in eastern Machakos District, which is a leading cotton producing area, and Kigumo Division, an area of high agricultural potential in Murang'a District about 90 km north of Nairobi, which is a major smallholder coffee-growing area. Makueni is semi-arid, with maize, pigeon peas, cowpeas, and beans as the main food crops, though there have been some recent changes in the crop spectrum. Cotton is an extremely important cash crop for most farmers (Machakos District Cooperative Union, 1983). Land is relatively plentiful, with many households owning ten hectares or more, although only about half of this is usually cultivated because of labor constraints. Despite the generally unfavorable conditions, Makueni is a food surplus region in most years, but nonfarm employment is limited by its remoteness. The Machakos Integrated Development Programme (MIDP), supported by EC donors, provides extensive infrastructural support to the Ministry of Agriculture and the local cooperative societies. The latter buy the cotton and some of the food crops produced by the farmers and supply them with most of their farm inputs, including pesticides and spraying equipment.

Kigumo is a highland area, rising from around 1200m in the east to about 2300m at the edge of the Aberdare Forest in the west, with rainfall of about 1200mm in the east and over 1800mm in the west. The characteristic landscape consists of steep, closely spaced ridges running west to east, separated by rivers in the valley bottoms. The land is densely populated, (Kenya, Ministry of Finance and Economic Planning, 1981) with small land holdings averaging under two hectares per household. Maize, beans, potatoes, and a variety of vegetables are the main food crops, and most farmers also grow coffee (having between 150-2000 trees), and tea in the upper areas. Fertilizer and pesticides are obtained on credit from the coffee cooperative societies, which also purchase farmers' coffee, or from local shops. The infrastructure is well developed, and many farmers have a nonfarm income. There is also extensive use of hired labor in agriculture, and, especially during the coffee harvest season, a regular migration of farm workers between the lower and the upper areas.

PEST HAZARDS AND RESPONSES

To improve pest management and mitigate the effects of pest losses among small-scale farmers, it is important to examine the role of these losses in the farming system and to identify the kinds of loss that are most in need of outside assistance and the type of assistance that would be most effective. The long term impacts of such losses were variable and could be summarized as follows:

i)   Significant impacts of pests and diseases on individual crops occurred in each of the farming systems;

ii) Pests and diseases were not the main problems faced by farmers. In Makueni, drought was the predominant hazard and in Kigumo, maintenance of soil fertility was the main farming problem, with market fluctuations also creating uncertainty;

iii) The impacts of pests depend on the interaction between the natural hazard and the human response, and both must be examined to assess the feasibility of reducing such impacts.

In our surveys, the impacts of pests and diseases on some crops were marked, causing them to decline or disappear from the farming system. The introduction and diffusion of other crops was being impeded by the inability to control pest losses. Most of the main crops currently grown can be divided into two sets: (a) those for which a high potential loss results in a vigorous response by farmers, the losses being controlled, but often at substantial expense; (b) a lower potential loss stimulates a lower level of response, often confined to extending the range of planting.

Tables 15.1A—15.1E classify the main crops grown now and in the past according to pest hazard and response levels.

TABLE 15.1 A

Factors influencing crop losses and farmers' responses in Makueni and Kigumo: i) main pests, ii) other factors affecting crop production and iii) farmers responses: Abandoned or Vestigial Crops

Makueni

Bulrush, finger millet	i) Birds can cause regular large scale damage. ii) Expansion of primary education removed child labor for scaring birds; little national market; taste preference for maize; prohibition of home brewing. iii) Reduction in planting caused concentration of damage; crops almost disappeared from area.
Cassava, sweet potatoes	i) Squirrels, pigs, porcupines, termites, disease. ii) Drought losses, end of mandatory growing with Independence; change in tastes. iii) Gradual reduction; little now grown; some still sold in local markets.
Green grams, black grams	i) Aphids; bollworms & other caterpillars. ii) Drought losses; little market demand; taste changes.
Cotton	i) Stainers (& other pests). ii) Attempted introduction by colonial administration; no pesticides available. iii) Abandoned after a few seasons.

Kigumo

Sorghum, bulrush millet	i) Birds. ii) Taste preference for maize; little market demand. iii) Millet abandoned first; bird damage concentrated on sorghum; mostly disappeared by Independence.
Pigeon peas	i) Caterpillars, diseases, aphids, birds. ii) Spread of smallholder coffee and intensification of land use made other crops more valuable. iii) Bean growing increased to replace peas in diet; peas still sold on local markets (imported from other areas).
Cowpeas	i) Aphids, diseases (unspecified). ii) Mainly women's crop. iii) Largely replaced by beans.
Sweet potatoes, yams, cassava, arrowroot	i) Various animal pests, diseases. ii) End of mandatory planting; displacement by English potatoes. iii) Gradual decline; some still grown & sold on local market; low seed availability.

TABLE 15.1 B

Factors influencing crop losses and farmers' responses in Makueni and Kigumo: i) main pests, ii) other factors affecting crop production and iii) farmers responses: Threatened Crops — Significant Declines or Impeded Introduction

Makueni	
Sorghum	i) Birds. ii) Primary education, limited national market; maize preference. iii) Large reduction in planting; small amount still grown, mostly near homes; new varieties developed to reduce bird damage; efforts to revive national market.
Cowpeas	i) Aphids, extremely high losses of peas, seed availability sharply declined. ii) Mainly women's crop. iii) Reduced planting; some use of cotton pesticides, but concerns over safety especially for leaf consumption.
Citrus	i) Scale insects, aphids, other pests and diseases. ii) Drought losses; little availability of pesticides or fertilizers; poor marketing infrastructure. iii) Some farmers attempting to grow trees, but often frustrated; occasional use of cotton pesticides.

Crops in Tables 15.1A and B have declined significantly or been abandoned because of pest losses, but other factors, economic, institutional, and cultural have also been critical. For sorghum and millets it was not in the first instance an increase in bird damage that led to the decline of these valuable, drought-tolerant crops, but rather a breakdown in the traditional system for controlling these pests by posting children in the fields. This was caused by the increase in primary education in rural Kenya following Independence in 1963. At the same time, the change in tastes in favor of maize, the lack of a national market for sorghum and millet, and the prohibition on brewing traditional millet and sorghum-based alcoholic beverages all reduced the incentives to grow these crops. Once the decline had begun, pest damage became a significant and probably dominant factor as the birds concentrated on the remaining stands of the crops.

Pigeon peas were widely grown in Murang'a as late as the 1960s (Kenya Colony, 1962) and are still sold in local markets. Outbreaks of diseases and caterpillar pests (perhaps American bollworm) were reported by farmers to have been principally responsible for the crop's disappearance in the Kigumo area, but the rapid growth in the intensity of land use and the spread of smallholder coffee growing in the 1960s also made it increasingly uneconomic to grow this two-season crop.

TABLE 15.1 C

Factors influencing crop losses and farmers' responses in Makueni and Kigumo: i) main pests, ii) other factors affecting crop production and iii) farmers responses: High Potential Threats, Substantial Response

Makueni

Cotton	i) American bollworms, stainers, mites, other pests. ii) High infrastructural support for marketing and supply of pesticides and other inputs (on credit). iii) Extensive pesticide use; some cultural controls (e.g. uprooting stalks).
Maize	i) Wild pigs (some areas), can cause massive destruction. ii) Main staple. iii) Fields guarded constantly as crop nears maturity.
Post-harvest storage (maize and other crops)	i) Weevils, other storage pests. ii) Sale of surplus food crops common. iii) Use of storage chemicals, ashes, chilli powder, smoking, repellent plants on stored food crops.

Kigumo

Coffee	i) Coffee berry disease (CBD) leaf miner, antestia, leaf rust, other pests and diseases. ii) Widely grown; high infrastructural support; major cash source. iii) Extensive use of fungicides and insecticides; some cultural controls.
Potatoes	i) Fungal disease (blight, etc.), bacterial diseases (wilt, etc.). ii) Major new food crop throughout country; local and national market; grown mostly for home use, also for sale. iii) Extensive spraying with coffee chemicals and other fungicides; attempts to plant resistant varieties from other areas.
Maize	i) Stem borers. ii) Land use intensification and reduced availability of land; wide availability of chemicals. iii) Extensive use of stem borer dusts.
Post-harvest storage	i) Weevils and other pests. ii) Food purchases common; little surplus grown. iii) Extensive use of storage chemicals; little use of ashes.

TABLE 15.1 D

Factors influencing crop losses and farmers' responses in Makueni and Kigumo: i) main pests, ii) other factors affecting crop production and iii) farmers responses: Significant Threat — Moderate or Low Response

Makueni	
Maize	i) Stem borers considered main maize pest by most farmers. ii) Drought losses predominate; no land constraint; pesticide available in societies but only in some shops. iii) Low rate of maize pesticide use; few cultural or traditional controls; perhaps planting extended to compensate for losses.
Pigeon peas	i) Pod borers, American bollworm, other pests and diseases. ii) Extensively grown for home use and sale; drought tolerant crop. iii) Some spraying with cotton chemicals; use of traditional controls (burning dung in field (rare)).
Beans	i) Pod borers, other pests. ii) Drought losses predominate; grown for both home use and sale; mainly women's crop. iii) Little pesticide use or other controls; perhaps increased planting.
Kigumo	
Beans	i) Pod borers, other pests & diseases. ii) Excess moisture often lowers yields; mainly women's crop, for home use; some sale. iii) Little pesticide use or other control; bought at local markets.

TABLE 15.1 E

Factors influencing crop losses and farmers' responses in Makueni and Kigumo: i) main pests, ii) other factors affecting crop production and iii) farmers responses: Low Pest Threat

Kigumo	
Tea	ii) Excellent infrastructure; low input requirements; harvesting throughout year with regular payments.
Bananas	ii) Widely grown throughout lower area for home use and sale.

Cultivation of the traditional root crops, cassava, sweet potatoes and yams, has been affected by animal pests and diseases and also by changes in tastes. During the colonial period, farmers were often compelled to grow cassava as a famine reserve in the event of locust attacks. Such legal requirements were eliminated with independence, and the absence of serious locust attacks in recent decades reduced farmers' perceived needs for these root crops. At the same time potatoes have become a major crop in Kigumo, taking the place of traditional roots and tubers. On maize, which is a far more important staple than these root crops, farmers respond vigorously to animal pests such as wild pigs or hippos.

Many of the abandoned crops have traditionally helped supply insurance against drought or other risks, and their removal increases the vulnerability of the cropping system. Government and internationally sponsored projects are trying to revive some of these crops, but such efforts are unlikely to be successful when the pest threats, and the other factors that induced farmers to abandon the crops, remain.

The importance of wild animal and bird pests in these farming systems was notable. These are frequently overlooked by entomologists and agricultural researchers, despite being regarded as serious threats in many parts of Africa and Asia (Mascarenhas, 1971; Heong, 1984). Research that focuses exclusively on arthropod pests may have limited practical application where these are overshadowed by animal pests. For example, although sorghum shoot-fly can cause significant yield reduction, bird pests and other factors so dominated farmers' evaluations that improved control of shoot-fly is not likely to have much impact on trends in sorghum cultivation.

Even when pest losses are potentially severe, farmers often continue growing crops that are important to them, either by expanding the area planted and absorbing the losses, or by increasing the intensity of pest control. The latter response is particularly characteristic of important cash crops such as cotton and coffee. These not only have large pest complexes, but a much lower tolerance of qualitative damage than on food crops. The price for second grade cotton, for example, is half that for the first grade, and lower quality coffee receives considerably reduced prices or is rejected at the factory. These requirements make it impossible to grow either cotton or coffee without the use of pesticides. Although analogous qualitative damage occurs on food crops, farmers can cope more flexibly. Damaged grain can be sold at local markets at a reduced price and even very damaged maize can be used for animal feed.

The extent and nature of potential pest losses also play an important role in determining farmers' actions. Stem borers generally cause predictable losses and although farmers in Makueni say these stem borers are their most destructive pests on maize (with losses estimated at 20-40%), they usually take no action against these pests, except perhaps to plant a larger area. Only about 20% of the farmers used the inexpensive insecticide dusts available to reduce stem borer infestations. By contrast, maize is occasionally attacked by wild pigs and other animals, in parts of Makueni, which can cause large losses, and then farmers guard their fields at night for extended periods. Interestingly, these farmers perceive stem borers as being of minor importance.

Unlike Makueni, in Kigumo 76% of the farmers were using stem borer dusts, which are inexpensive and readily available in shops and coffee cooperative societies. A reason for this difference in approach may be land availability. Farmers in Kigumo often have barely enough land to grow maize and other subsistence crops, and almost all used commercial fertilizers on their maize. Minimizing stem borer losses is thus an understandable priority. In Makueni, land constraints are not significant, and the availability of family labor is the main factor limiting the amount of land cultivated. Thus, increasing the area of maize planted can compensate for stem borer and other predictable pest losses.

In Makueni there was, however, widespread use of pesticides in storage, perhaps due to the nature of the damage involved. Losses to weevils and other storage pests are irretrievable in a sense that is not true of field pest losses, and they are inflicted after considerable investment in the crop has already been made. Also, farmers say that if they do nothing, weevils will consume all of their stored maize, while stem borers and other field pests, even without control measures, will destroy only a portion of the crop.

The use of pesticides for cash crops had some important spill-over effects for pesticide use on food crops. In both survey areas, the availability of chemicals and spraying equipment, the presence of the cooperative societies which provide credit for purchase of these, and farmers' general familiarity with pesticide use have led to increasing use of pesticides on a number of food crops. For example, in Kigumo 70% of the farmers surveyed were growing potatoes, mainly for household consumption, and of these 90% used fungicides, a rate of pesticide usage comparable to that on coffee and cotton. These farmers generally sprayed potatoes and coffee at the same time to save on such spraying costs as water, hired labor, and rented equipment. Similarly, in Makueni, cotton chemicals were used on pigeon peas and cowpeas. Thus, the diffusion of pesticide use on cash crops has had significant impacts on food crops, in some cases permitting the continued cultivation of crops threatened by pests and diseases.

ASPECTS OF PESTICIDE USE

Neither cotton nor coffee can be grown without pesticides, and since these crops are sold to parastatal marketing organizations, farmers are supplied with credit for chemicals and other inputs, charging these against earnings on the delivered product. In Makueni, cotton pesticides and spraying equipment were often the only inputs purchased by farmers. All the cotton farmers surveyed used pesticides in the 1981/82 season, at a cost of just under 10% of the gross income from cotton. Coffee farmers, in comparison, used a much wider range of inputs, including fungicides, insecticides, hybrid maize seeds, and, the largest input, fertilizers, the latter being used on coffee and various food crops. (In one cooperative society, fertilizers accounted for over 80% of purchased inputs in the 1979/80 season, compared to 16% for all pesticides.)

Availability and Selection of Pesticides

The pesticides available for sale to farmers must be on the list of those tested and approved for the intended use by the National Agricultural Laboratories (NAL). The specific chemicals supplied are determined by negotiations between the district cooperative union and chemical companies, taking account of cost, safety, and appropriateness for local conditions. Changes in availability of chemicals occur as a result of changes in government policies, pest-related conditions, or individual transactions between the societies and chemical suppliers, but farmers are often confused about the reasons for such changes. Coffee farmers and societies were better informed, had more experience, and exercised more choice in selection of pesticides than cotton farmers and societies.

Cotton farmers generally used only the one main insecticide sold by their society, until 1980 a combination of DDT and carbaryl. After 1980, the synthetic pyrethroids permethrin and cypermethrin, in a variety of formulations (including ultra-low volume [ULV]), were the main insecticides supplied. They raised the costs of insect control, but they were effective and farmers were generally satisfied with them. They were also perceived as relatively safe, and were used on some food crops.

However, spider mite (*Tetranychus telarius*) became an increasing problem on cotton, perhaps because the mites' predators had been eliminated by the broad spectrum pyrethroids. Special miticides were made available but farmers generally did not recognize the problem, and very few purchased these additional chemicals. Consequently, a new pesticide, combining cypermethrin and the organophosphate miticide profenofos, was supplied in 1982/83. This further increased costs to farmers, and also caused safety concerns since it had a strong odor and farmers frequently felt ill when using it. Farmers were reluctant to use it, and, combined with the fact that little information was provided on recognizing and combatting the mites, this has meant that efforts to control them have met with limited success.

A wider range of chemicals is available to farmers in the coffee growing areas of Kigumo. Fungal diseases, particularly coffee berry disease (CBD) and, to a lesser extent, leaf rust are the main problems. A variety of insect pests also need to be controlled, including leaf miner (*Leucoptera* spp.), thrips, and various scale insects. Copper fungicides have been used for many years and resistance and phytotoxicity may now be developing. Organic fungicides are also being used, particularly captafol and dithianon, but these cost four to five times more per application than copper compounds. Use of captafol is sometimes avoided because of its hazards to health. Insecticides are used less than fungicides, and only one or two kinds are usually supplied. Resistance and other problems have caused changes away from DDT and malathion to the organophosphates, fenthion and fenitrothion.

Herbicide use (particularly paraquat) has also begun to increase, stimulated by increases in the costs of hired labor. Problems with a number of perennial grasses and sedges are also leading to increased use of the systemic herbicide glyphosate. Twenty seven percent of the farmers surveyed used herbicides in 1981/82, mostly growers with 1900 trees or more (approx 1 ha). Given trends in the cost of labor, herbicide use among smaller farmers is likely to grow.

Pesticide Spraying Behavior

Frequency, dosage, and timing of pesticide spraying (behavioral variables that are key elements of integrated pest management strategies) are subject to a variety of influences among cotton and coffee farmers. Economic and institutional factors, official recommendations, and farmers' perceptions and assessments are all involved. The data from these surveys show some significant differences between cotton and coffee farmers and raise questions about the role and impact of official recommendations.

Cotton. Cotton farmers are recommended to spray five times per year, especially at flowering and other critical points (Kenya, Ministry of Agriculture [MOA], 1982). This advice is passed on by extension agents at the local level. A parallel set of recommendations for pest scouting exists (MOA, 1972) but this advice only reaches the few farmers who attend Farmer Training Center (FTC) courses, and is rarely conveyed by extension agents in the field.

The mean number of insecticide sprays used by the farmers surveyed was 4.4 in the 1981/82 season. The majority spray when they see pests or signs of their effects, particularly American bollworm and stainers, and they are aware of the importance of spraying at key points. Formal scouting is rare and instead farmers employ techniques and rules of thumb that they develop themselves or learn through informal networks such as neighbors, friends, or other farmers. Spray timing is adjusted to climatic conditions and expectations of yield. For instance, despite the buildup of aphids and other pests during the dry season between December and March, most farmers know that it is not worth spraying at this time. The March rains are known to be unpredictable in onset and duration and spraying is usually delayed until the rains begin and is then closely geared to yield expectation. Only a few farmers follow the type of calendar schedule recommended by most extension agents and chemical companies.

The survey results also indicated a striking divergence from formal dosage recommendations, mean dosage per spray being about half the recommended amount. Apart from one pyrethroid formulation that was less concentrated than the others, this was probably because farmers were spraying less thoroughly per unit area than advised, to economize on chemical usage, rather than mixing the chemicals incorrectly. The suitability of standard dosage advice in terms of the amount of concentrate to be used per cropped area is also subject to question. Farmers generally do not know accurately the area planted to cotton, and variations in plant density and pest infestation, different stages of the crop's life cycle, and climatic conditions can affect the appropriate intensity of pesticide application. Farmers may be responding to these variable conditions, under the general pressure of economic constraints, rather than making mistakes.

Coffee. Coffee farmers also time their spraying according to pest sightings and observations of damage and spray mostly during the colder rainy periods in March, April, and June when conditions were conducive to CBD. They also tended to combine insecticide and fungicide sprays because of the high fixed costs of each application. (Potatoes and other food crops were often sprayed at the same time for these reasons of economy.) CBD generally takes precedence over pests and governs the timing of combined sprays.

Farmers are strongly recommended to spray nine times a year, on a calendar schedule, for CBD control (Coffee Research Foundation, [CRF] 1983). Recent findings indicate that less frequent spraying may be counter-productive by eliminating saprophytic fungi that compete with CBD (Masaba, CRF, pers. comm., 1983). On the other hand, the latest recommendations for insect control emphasize pest scouting rather than calendar spraying (CRF, 1980a; b), using scouting techniques being developed for smallholders and plantations (Bardner, 1979). However, most farmers continue to use their own informal methods of scouting and deciding on spray timing. Faced with two opposite sets of recommendations, calendar spraying for CBD and scouting for insects, there is an understandable tendency for farmers to continue operating according to their own observations.

The economic pressures on coffee growers have been increasing. The costs of pesticides and fertilizers have risen 1.5 to 2.5 times over the last five years, while the prices paid for coffee have declined almost 30% between 1976/77 and 1982/83. As a result, coffee farmers look for ways to improve coffee profits or minimize losses. The outcome is a wide range of spraying behavior and general agronomic practices, as farmers search for appropriate strategies. The mean number of fungicide sprays by farmers in our sample was 4.9, considerably less than the recommended nine per year, although there was a wide distribution from zero to more than 12. However dosages used were often considerably more than recommended especially for the less expensive copper fungicides, but also for captafol and dithianon, to compensate for the reduced frequency of spraying. Such behavior may make economic sense to farmers since the fixed costs of each application far exceed the cost of the additional chemicals used when over-dosing the crop.

In general, it can be said that cotton and coffee farmers have two distinct pesticide strategies which are used under different conditions. A 'basic strategy' is employed when crop prospects are not good, because of climatic conditions (particularly on cotton) or price expectations, and only the minimal number of pesticide applications is used at the times the farmer judges to be most critical. A 'de luxe' strategy is used when prospects are particularly good, for example when the price of the crop and/or the growing conditions are favorable, involving a higher and more frequent use of pesticides, approaching, and in some cases exceeding, the ideal spray regime officially promoted. The advice farmers receive and most of the research on pest management related only to the latter approach, but in most instances, farmers employ a 'basic' strategy. They would benefit from advice on undertaking a suboptimal strategy when this is warranted. In other words, farmers need to know not just the optimal approach to pest management, but also how best to depart from this optimum when necessary. There is little indication that researchers or extension personnel are attempting to develop and promote such a flexible approach.

THE ECONOMICS OF PESTICIDE USE

A number of general observations can be made about the economics of cotton and coffee in the areas studied which have implications for pest control behavior.

Cotton as a cash crop offers very low returns in Kenya, because of the low official price, the high labor requirements and the need for chemical inputs. Economic studies indicate that net returns are comparable with the wage rates for casual labor (Heyer, 1967; ICRA, 1982 and 1983). In Makueni, however, cotton is widely and seemingly successfully grown, with consistently rising production in recent years, despite the low rainfall and difficult growing conditions. Most of the farmers surveyed claimed to be growing cotton because of its high economic returns.

This disparity in assessment seems to be due to a number of factors. Alternate opportunities for earning cash are limited in this fairly remote area. Even if earnings from cotton are not much greater than casual labor rates, little hired labor is used in the area, and the opportunity cost of labor is low. The net amounts earned from cotton ranged from KShs 1,250 to 4,500 (approximately US $125 to $450 at the 1982 exchange rate), the largest single source of agricultural income for most farmers. Unlike income earned from food crops such as maize and pigeon peas, the cotton payment comes in one lump sum which gives farmers the opportunity for investment or consumption purchases they otherwise rarely have. Also, unlike food crop sales, the price for cotton is known in advance and does not fluctuate with local conditions; in particular, it does not decline if there is an especially good harvest. Finally, cotton growing is principally a male activity, and in its absence, there would be higher pressures for men to migrate to urban areas in search of employment. The value of pesticides used by farmers was about 10% of gross earnings. This was the main financial input, commercial fertilizers not being used in Makueni. Farmers thus felt that this was the main area where they could make savings when growing conditions were unfavorable or the price for cotton was low.

The situation of coffee farmers was more complex than that of cotton farmers. A substantial proportion felt that coffee production had become so unfavorable that it represented a net economic loss, and this affected their decisions on pesticide use. An economic analysis shows that this is true only at low input levels and scales of production, (Goldman, 1986). However, few farmers were ready to give up coffee growing because, to an even greater extent than for cotton, the profit-loss relationship expresses only one of a number of relevant economic aspects of production. Five key factors mentioned by farmers were:

i)   *Credit Availability.* The credit for input purchases obtained from cooperative societies is one of the main benefits from coffee production. Fertilizers, purchased on credit, are used for maize and other food crops, yields of which would otherwise be extremely low. Cash loans are also provided and used for coffee or other agricultural or nonagricultural investments. Credits and loans are, however, contingent on coffee production and are a prime incentive for farmers to continue growing coffee and maintaining reasonable output levels.

ii) *Windfall Profits.* The coffee boom in 1976/77 generated windfall profits for farmers and convinced many to go into coffee production or to expand their acreage. Worldwide increases in coffee planting make a recurrence of this boom unlikely, but farmers still hope for another windfall season and many believe that the possibility of windfall gains makes coffee a more profitable crop in the long run than tea, which provides a more predictable and stable economic return.

iii) *Price Uncertainties.* Calculating the benefit of inputs on coffee is impeded by uncertainty about the final price to the farmer. Each factory pays its members a different rate, depending on its running expenses and the overall quality of its production. In 1980/81, there was more than 100% difference in the prices paid to farmers between two of the societies in Kigumo Division (Kenya, Ministry of Agriculture, 1981). There was thus little incentive for a farmer to produce higher quality coffee than the mean at his factory, creating downward pressure on the level of input use and the quality of coffee production.

iv) *Factory Labor Requirements.* Most processing labor at each factory is supplied by member farmers, 100-150 person days per household per year, regardless of the amount of coffee it produces. Much of this labor requirement coincides with food crop harvests and is very costly to the household. It can make coffee production prohibitive for smaller growers, especially in periods of low profit margins.

v)   *Regulation.* Coffee is a protected crop and maintenance of the high quality of Kenyan coffee is a major national policy. It is illegal to uproot trees without permission, and practices such as intercropping coffee with beans (becoming common as the profitability of coffee declined) are prohibited to avoid general deterioration in quality.

Given the complexity of this general economic situation and the marginality of profits, decisions on the optimal use of pesticides are difficult. Clearly the economically relevant questions go beyond determination of the action threshold, which has been the main focus of research based on Western agricultural conditions. In contrast to the latter, underuse of pesticides is the norm, and the main economic question is how to achieve the best net return given the numerous constraints. No single strategy is likely to produce optimal results for all farmers. The objective of research and extension should be to tell farmers how to determine the strategy that will produce the best results in their particular situation. This is a more complex and demanding goal, but one that may produce appreciably more successful results in the long run.

REFERENCES

Bardner, K. (1979). Integrated Control of Coffee Pests. Ruiru: Coffee Research Foundation [CRF] (mimeo).

Coffee Research Foundation (1980a). Control of coffee leaf miner. Technical Circular No. 27. Ruiru: CRF (mimeo).

ibid (1980b). The antestia bug: damage, testing and control. Technical Circular No. 32. Ruiru: CRF (mimeo).

ibid (1983). Control of coffee berry disease and leaf rust in 1983. Technical Circular No. 54. Ruiru: CRF (mimeo).

Goldman, A. (1986). Pest hazards and pest management by small scale farmers in Kenya. Ph.D. dissertation. Clark University USA.

Heong, K.L. (1984). Pest control practices of rice farmers in Tanjong Karang, Malaysia. *Insect Science and its Application* 5, 3: 221-226.

Heyer, J. (1967). *The Economics of Small-Scale Farming in Lowland Machakos*. Nairobi: Institute of Development Studies. Occasional Paper No. 1.

ICRA (1982). The farming system in Makueni location, Machakos, Kenya. Wageningen, Netherlands: ICRA (International Course for Development Oriented Research in Agriculture). Bulletin No. 8.

ibid (1983). Agricultural development and research priorities for a semi-arid area of Machakos district, Kenya. Wageningen: ICRA. Bulletin No. 10.

Kenya Colony (1962). *Fort Hall District Gazeteer.*

Ministry of Agriculture, Kenya (1972). *Cotton Growing Recommendations for Kenya.*

ibid (1981). *Murang'a District Annual Agricultural Report.*

ibid (1982). *Pesticide use recommendations for cotton.*

Ministry of Economic Planning and Development, Kenya (1981). *Kenya Population Census, 1979, Vol. 1.*

Machakos District Cooperative Union (1983). (District cotton production figures).

Mascarenhas, A. (1971). Agricultural vermin in Tanzania. In *Studies in East African Geography and Development*, ed. S.H. Ominde, pp. 259-267. Berkeley: Univ. of California Press.

16

The Environmental Conflict And Farmers' Attitudes To Pesticide Use In Britain

S. Carr

INTRODUCTION

Although fewer than 3% of Britain's adult population now work on the land, they produce more than 75% of the country's needs in temperate food (HMSO, 1983). Successive governments since 1945 have encouraged farmers to increase their productivity and have supported this with: guaranteed prices; grants towards farm improvements; the Agricultural Development and Advisory Service (ADAS); funding for agricultural research and training; and tax incentives. A government white paper on agriculture (HMSO, 1979) pledged continuing support for this policy by concluding "..import prospects and the need for insurance continues to point to the desirability of increased agricultural output in this country."

Agricultural production has been further encouraged since 1973 by Britain's entry into the European Community (EC), the stated aims of the EC Common Agricultural Policy (CAP) being to: increase agricultural productivity; ensure a fair standard of living; stabilize markets; guarantee regular supplies; and ensure reasonable prices (National Farmers Union (NFU), 1971).

With increasing agricultural production, the level of concern about its impact on the rural environment has also increased. Intensive agriculture is claimed to be impoverishing the country's landscape (Countryside Commission, 1974), its wildlife (Nature Conservancy Council, 1977) and natural resources in general (O'Riordan, 1982). Support for the environmental movement is manifested in the rapid increase in membership of the Royal Society for the Protection of Birds (RSPB), one of the largest and most influential conservation groups. Their membership increased from below 10,000 in the 1940s and 1950s to 65,000 during the 1960s and more than 380,000 in 1984. Conservation organizations in Britain now claim a combined membership of more than three million people, far outnumbering the 300,000 farmers.

Environmental groups accuse farmers of 'vandalising' ancient woods, destroying hedgerows, enveloping the countryside in a 'pall of poison' with pesticides, threatening the existence of wetland wildlife by drainage schemes, setting light to the countryside by burning cereal straw, preventing public access to farmland and, more recently, of burdening the taxpayer with the cost of the EC's surplus food production.

The dispute between agriculture and conservation is generally seen as one of conflicting interests and competing uses for the scarce resource of land. The government policy of designating certain areas where conservation takes priority over agricultural production, and compensating the farmer for the resulting financial loss (HMSO, 1981), reflects this understanding of the problem. But the intense and often emotional nature of the accusations suggests that conflicting sets of values are also involved, conservationists and farmers each viewing the problem in a different light. In such a case, the conflict is unlikely to be resolved by monetary compensation alone.

To gain a better understanding of the nature of this conflict, surveys of the attitudes of (a) local farmers and (b) members of organizations with an interest in the countryside ('conservationists'), to farming and conservation have been carried out in Bedfordshire. This county lies between the intensive cereal growing area of East Anglia and the mixed agriculture of the Midlands.

An unstructured survey was carried out to establish what aspects of conservation concern people and why. The topics most often mentioned were loss of hedges and trees, pesticide and fertilizer use, loss of marshy areas, ponds, ancient woods and wildlife habitat in general, straw burning, access to the countryside and surplus food production. Conservationists were almost all concerned about the use of pesticides and also about the loss of hedgerows, trees and other wildlife habitat. On the other hand farmers did not appear to consider pesticides a conservation-related issue; in general their main conservation concerns were about public access to farmland, the possibility of a ban on straw-burning and other such restrictions, although they shared some of the conservationists' concern about the loss of trees.

This paper describes the theoretical framework of the attitude survey, the Ajzen and Fishbein theory of reasoned action, and discusses some of the results of an unstructured survey with particular reference to pesticides.

SURVEY METHODS

The theory of reasoned action (Ajzen and Fishbein, 1980) relates attitudes to behavior and has been used in a variety of practical situations, including surveys of farmers' attitudes to pesticide use (Tait, 1983). It takes account of underlying beliefs, values and social influences, all of which are important to an understanding of conflicts over conservation and agriculture. The model is summarized as a linear regression equation:

$$B \approx BI = (A_B)W_0 + (SN)W_1$$

where B is behavior, BI behavioral intention, A_B attitude towards the behavior, SN subjective norm (social influences) and W_0 and W_1

empirically determined weights.

A preliminary survey was done to establish relevant beliefs, values and social influences and to develop an index of conservation behavior for the model. Twenty four farmers and 26 members of local organizations with a countryside interest were interviewed. Five farmers were county committee members of the NFU or the Bedfordshire Farming and Wildlife Advisory Group (FWAG) and the others (all with holdings over 20 ha) were randomly selected from the NFU membership list (claimed to include 93% of Bedfordshire farmers). The countryside interest groups (and their local membership) were as follows: the Naturalists Trust (4,000), the Natural History Society (400), the Royal Society for the Protection of Birds (4,000), the Preservation Society (1,000), the Ramblers Association (160), Friends of the Earth (40) and the Conservation Volunteers (30-40). Those interviewed were either committee members or randomly selected from membership lists.

The interviews took the form of unstructured conversations around a common core of questions about conservation and farming. Specific questions were asked only if respondents omitted to mention topics of interest to the research. The tape-recorded interviews were transcribed verbatim and content-analyzed (Whyte, 1977), recurring themes, sentences and phrases of interest being selected and grouped together. These were used to construct a questionnaire based on the Ajzen-Fishbein model for a second survey, and to assess whether this model could encompass all the attitude dimensions in a complex issue like conservation.

RESULTS

Pesticides, and agrochemicals in general, were frequently mentioned by conservationists as one of their main concerns about farming. Also, in a recent public opinion poll (MORI, 1983) 10% of those interviewed saw agrochemical use as the main threat to the countryside. Farmers, however, did not see pesticides as a conservation-related issue and only mentioned them in response to a specific question.

Attitudes

Farmers generally felt that pesticides were an essential part of modern farming and that their safety, to people and the environment, was carefully tested by scientists and manufacturers (Table 16.1). Although mistakes had been made in the past, for example with arsenical compounds and DDT, they felt that pesticides were now safer, used at much lower concentrations and applied with better equipment. Fungicides in particular were routinely used and considered harmless. Cost seemed to be the only disincentive to the use of insecticides and herbicides in a similarly liberal way.

Only four conservationists were not concerned about pesticide use: three of these had close farming connections and one had formerly worked

TABLE 16.1
Beliefs about pesticide use

	The use of pesticides:
Mainly farmer beliefs	is an essential part of high input systems
	ensures high yields
	allows us to keep on top of pests, diseases and weeds so they don't build up
	is now restricted to carefully tested chemicals
Mainly conservationist beliefs	harms wildlife
	leads to a build-up of pesticides in the food chain
Beliefs shared by some farmers and conservationists	provokes worse strains of pest and disease
	harms beneficial insects
	leaves toxic residues in the soil, water or crop
	makes us over-dependent on chemicals
	affects health

for a pesticide company. Most were less certain than farmers that the testing of pesticides was adequate, and they were worried about their unknown long-term effects. A typical remark was, "O.K. well presumably they're safe but I don't know. I don't know how long it took before the chloro-bi-phenols caused the decline of the birds of prey, the thinning of their egg-shells ... there's no telling just what long-term effects there might be from all sorts of chemicals that are being used now." (RSPB member).

Conservationists' main concern was about harmful effects of pesticides on wildlife, both wild flowers and animals, particularly the risk to animals at the top of the food chain, whereas farmers generally felt the newer pesticides were relatively harmless in this respect (Table 16.1).

Most conservationists felt that the amounts of pesticide used were too small to affect human health. This relatively low level of concern, compared with that in less developed countries (Bull, 1982), probably reflects the small likelihood of most people coming into contact with pesticides, other than those intended for garden use. Some farmers mentioned mishaps in handling pesticides, but only two people, a farmer's wife and a vegetable grower, expressed concern about their effects on health.

A sizeable minority of farmers did express some reservations, particularly about insecticides, and some avoided using them when possible. One farmer expressed strong reservations about pesticides, and yet as a vegetable grower, felt it necessary to use them frequently. Similar cases where pesticide use conflicted with attitude have been noted by Tait (1982).

About half of the conservationists who were worried about pesticide use qualified their remarks by saying that, from the farmers' point of view, there were advantages but only two mentioned a plentiful supply of food as a benefit of pesticides from the public point of view. Increasing the supply of food was usually seen as a dubious benefit, given the surplus production in the EC, for example: "The thing that really gets me about it is they're going hand-over-fist trying to bring every field into arable cultivation, trying to produce more and more wheat and barley, and they're already over-producing, so we don't really need this stuff." (Conservation Volunteer).

The fact that few farmers saw pesticides as a conservation-related issue suggests, on the basis of the Ajzen-Fishbein model, either that they are unaware of any pressure to reduce their pesticide use or that the pressure is coming from sources whose opinions they do not take seriously. A further explanation, which seems likely from the findings of this survey, is that the pressures from respected sources are overwhelmingly in favor of maintaining or increasing pesticide use, for reasons other than conservation.

The survey showed that neighboring farmers and farmers' merchants can be a powerful influence, encouraging increased use of inputs in general. As one farmer said, "You try to be as good as your neighbor. You get the ICI boys coming round saying 'Your crop isn't looking so well, you want to use more nitrogen'." (Arable farmer). ADAS was a respected source of advice, "I take a lot of advice from ADAS because they're the impartial ones." (Arable farmer). However, there was no suggestion from the farmers that ADAS recommended reduced pesticide use as a conservation measure, although ADAS is now required to take conservation into account when giving advice. The fact that conservation is usually handled by the Land and Water Service Division of ADAS, while advice on pesticides and fertilizers is given by the Agricultural Service Division, may be partly responsible for this.

The only pressure on farmers to reduce pesticide use came from members of their own family, in one case a wife and in another (quoted here) a father: "Dad and I have quite a few arguments (about pesticides)... father's of the opinion that I would always use them which isn't necessarily so and I'm of the opinion that he'd never use them which isn't necessarily so." (Arable farmer).

Conservation organizations appear, from this survey, to exert no pressure on farmers' pesticide use. Given the strength of their feeling about pesticides, this is surprising. On the other hand the farming community had some influence over the conservationists, in that many of the latter qualified their statements of concern with an appreciation of the farmers' point of view. Conservationists' views were likely to be reinforced by their societies' journals and talks, and by watching natural history programs on television. "We watch all these wildlife programs on TV and it's amazing, you never get to the end without some pessimistic point of view coming over". (RSPB member).

The picture that emerged from this survey indicated that many farmers were isolated from the views of people outside the farming community. For example, "Farmers stay very remote from everybody... Most of my friends are involved in farming, and those that aren't have quite a lot to do with people that are". (Arable farmer). Farmers repeatedly mentioned how few people living in rural areas now worked on the land or had any understanding of farming, and felt this was one of the main reasons for the conflict: "The trouble is once all the people who lived in the village were people who worked on the farms or were in some way or another connected, but now there aren't the people working on the farm and the farmhouses and cottages are being sold off and townspeople come into them". (Mixed farmer).

Consequently there is a declining influence of farmers on their local society which may explain why the conflict has surfaced to its present extent: "At one time I think farmers were the leaders of the community, but of course the farmers don't lead the community now because they don't employ these people." (Arable farmer). This isolation has also been noted by Newby et al. (1978).

Behavioral Indicators of Concern about Pesticides and Conservation

Relevant behaviors mentioned by farmers were: not using pesticides routinely; avoiding harm to bees by not spraying crops in flower, using less toxic chemicals or spraying late in the day; avoiding contamination of streams; avoiding the use of pesticides too close to harvest; and washing carefully after their use.

For conservationists the most appropriate behavioral indicators seemed to be derived from the action they were prepared to take in order to influence farmers' pesticide use. This was very much a reflection of the different approaches used by the conservation organizations to which they belonged. Membership of the organization itself was implicitly assumed by

- 137 -

the survey to be one indicator of conservation behavior and many of those interviewed belonged to more than one group.

Only the two smallest groups encourage direct action: the Conservation Volunteers and Friends of the Earth. The Conservation Volunteers carry out conservation tasks on farms in response to requests from farmers. The approach of Friends of the Earth (FoE) is more confrontational; members might hold a demonstration on the farm to attract publicity through the press and embarrass the farmer. One FoE member had petitioned all his neighbors to write to the Department of the Environment and complain about frequent spraying of an adjacent field.

Members of the Naturalists Trust contribute money towards the purchase of farmland which is then managed for nature conservation. This approach is also used by the RSPB, although they concentrate on nationally important sites. Some conservationists felt the only way to bring about change was to educate people to their own way of thinking, by giving lectures, writing articles and contacting their MP: "Basically we cannot do anything other than attempt to educate; as a charity we're not allowed to have any political bias. Obviously we can talk to our MPs and our representatives on our local government but we have no power to take any action, we can only try and educate the people." (RSPB member).

The Preservation Societies work by cultivating good relations with local government and the local press, but are more concerned with local planning than the broader issues involved in farming and conservation.

The fact that larger conservation organizations favor indirect action such as the purchase of land or persuading people to support their point of view and influence government legislation, may explain why individual farmers perceive little pressure at the local level to reduce their pesticide use.

Conflicting Attitudes and Norms

Although they held differing beliefs, for example over the extent to which pesticides damage wildlife, some farmers and conservationists shared similar sets of values.

For these people some form of compromise, such as that already initiated by FWAG, should be a relatively simple matter. As one RSPB member said: "When you get to know and talk to people and really know what they believe and think, their ideas are not so far removed from the ideas that true conservationists may hold".

In other cases, particularly over perceptions of attractive landscape and of land ownership, farmers and conservationists saw the issues in a very different light. Typical of many farmers' views about landscape was, "To me it's attractive because it's neat, presentable and it looks tidy". (Arable farmer). Conservationists were more likely to say: "I don't like the way they've 'cleared up' the countryside. There's not enough rough and tumble that we used to have, nor the wildlife". (Naturalists Trust member).

On the question of who owns the land many farmers had a strong sense of ownership, for example, "I think it's dreadful if you own something... that somebody else should dictate you shouldn't do this, that and the other". (Livestock farmer). Whereas conservationists saw land as held in trust by farmers: "I think we ought to get over to them that they are just custodians of the land, they're holding it for our future generations". (Naturalist Trust member).

Cotgrove (1982) and Buss and Craik (1983) have discussed the tendency for people to develop compatible sets of attitudes to a range of contemporary issues. Where opposition crystallizes across such a broad front, this can be a particularly serious source of conflict, and there was some evidence of this taking place between the farmers and conservationists surveyed. For example one conservationist said: "I don't really know that much about it (conservation) but I do care about what's happening — I suppose I'm basically concerned about the society we live in, in general. The countryside is one side of it... My interest in nuclear power was first... people that tend to be anti-nuclear power tend to be conservation-type people, so then you get on to other topics when you're talking to them, for instance general resources". (Conservation Volunteer). The antagonistic view from a farmer was, "I'm anti-conservationists. Most of them that I've come across that would call themselves staunch conservationists drive me nuts ... They all strike me as left wing fanatics, or maybe right wing fanatics, but fanatical". (Arable farmer).

Statements based on sets of values rather than specific beliefs provoke an angry response from those with different opinions, who dismiss them as ignorant and irrational. The conflict is heightened by the minimal direct contact between farmers and conservationists, as it is often only those with such strongly held convictions who receive any attention, and they are dismissed as cranks. Such value-laden conflicts are difficult to resolve by negotiation and usually require legislative action.

CONCLUSION

Pressure from the environmental lobby, among other factors, has now caused the British government to rethink its agricultural policy and give more positive support to conservation. The two major farming unions, the National Farmers Union and the Country Landowners Association, have also agreed that there should be less emphasis on productivity and more on conservation. Recent British pesticide legislation, the Food and Environment Protection Act, 1985, includes powers to enable the government to control pesticide use at the farm level.

With government and the environmental lobby thus equipped for action, adverse publicity from an incident involving pesticides could completely alter the balance of pressures on farmers over their use. Given the apparent contribution of such pressures to pesticide usage behavior, and the fact that a sizeable minority of farmers have ambivalent attitudes to pesticides, an unexpectedly abrupt change to reduced pesticide use could take place. A smoother transition, more attuned to farming needs, might be achieved if the farming community were to take more notice of public opinion now and attach greater importance to precision in pesticide use.

ACKNOWLEDGMENTS

Special thanks are due to the Open University for funding the research and to Dr Joyce Tait whose ideas both initiated the research and continue to contribute to it.

REFERENCES

Ajzen, I. and Fishbein, M. (1980). *Understanding Attitudes and Predicting Social Behavior*. Englewood Cliffs, N.J., USA: Prentice Hall.

Bull, D. (1982). *A Growing Problem — Pesticides and the Third World Poor*. Oxford: Oxfam.

Buss, D.M. and Craik, K.H. (1983). Contemporary worldviews: personal and policy implications. *Journal of Applied Social Psychology* 13: 259-280.

Cotgrove, S. (1982). *Catastrophe or Cornucopia*. Chichester: John Wiley and Sons.

Countryside Commission (1974). *New Agricultural Landscapes*. Cheltenham, U.K: Countryside Commission.

HMSO (1979). *Farming and the Nation. Cmnd 7458*. London: HMSO.

HMSO (1981). *Wildlife and Countryside Act 1981. Chapter 69*. London: HMSO.

HMSO (1983). *Annual Review of Agriculture. Cmnd 8804*. London: HMSO.

MORI (1983). *Public Attitudes towards Farmers*. London: MORI.

Nature Conservancy Council (1977). *Nature Conservation and Agriculture*. London: NCC.

Newby, H., Bell, C., Rose, D. and Saunders, P. (1978). *Property, Paternalism and Power*. London: Hutchinson.

NFU (1971). *Farmers and Growers Guide to the EEC*. London: National Farmers Union.

O'Riordan, T. (1982). *Earth's Survival. A Conservation and Development Program for the U.K. Report No. 7. Putting Trust in The Countryside*. London: Kegan Page.

Tait, E.J. (1982). Farmers attitudes and crop protection decision making. In *Decision Making in the Practice of Crop Protection*. Proceedings 1982 British Crop Protection Symposium, 43-52 (Ed. R.B. Austin).

Croydon, England: British Crop Protection Council.

Tait, E.J. (1983). Pest control decision making on Brassica crops. *Advances in Applied Biology* 8: 121-188.

Whyte, A.V.T. (1977). *Guidelines for Field Studies in Environmental Perception. MAB Technical Notes 5.* Paris: United Nations Scientific and Cultural Organization.

17

Perception And Management Of Pests And Pesticides By Malaysian Traditional Small Farmers — A Case Study

Mohd Yusof Hussein

INTRODUCTION

More than 500,000 households can be classified as traditional small farms in Peninsular Malaysia. The average farm size is about two hectares with considerable variation between urban and rural areas, and among different communities in the different states. Almost two-thirds (65%) of these farmers are owner-occupiers, 24% are tenant farmers and 11% owner-tenant operators. They grow a large variety of crops such as tobacco, rice, fruits, vegetables, spices, cocoa, rubber, oil palm, coffee, cassava and yam, and some also keep livestock such as chickens, goats, cows and ducks (Ooi et al., 1983). Most traditional small farmers operate with limited means, with minimal use of modern technology and machinery, relying on family labour, hired labourers and sometimes community help.

Pesticide usage among traditional small farmers depends on the type of crop planted and other social and economic factors. Hardly any pesticides are needed on fruits, coconuts, cocoa and cassava (Ooi et al., 1983). On the other hand vegetable, tobacco and rice farmers use large quantities of pesticides, particularly insecticides. As in most developing countries, efforts by the government and associated agencies have not resulted in technical packages capable of increasing small farmer net returns through effective pest control. Critics complain that past programs lack understanding and appreciation of farmers' perceptions, needs and capabilities.

Agricultural development in Malaysia has concentrated on major plantation crops such as oil palm, cocoa and rubber, operated on a large scale by private corporate bodies. The government has only recently placed a high priority on improving the agricultural productivity and economic well-being of small farmers. In fact, a global concern is now demonstrated by international donors to focus on research and extension for small farmers, and Malaysia has formulated a new National Agricultural Policy aimed principally at small farmers.

The starting point for such programs should be an understanding of traditional agricultural systems and small farmers' perceptions, using them as the basis for future work. With the advent of integrated pest management (IPM), which is new to most small farmers, the process of decision making includes farmers' perceptions of crop damage caused by pests, attitudes to risk, perceptions of the cost-benefit ratio of control measures, and constraints on options and information (Norton, 1982a; Norton and Mumford, 1983).

The survey described here concentrates on traditional small farmers in Peninsular Malaysia, addressing particularly the prospects for IPM extension. Emphasis is given to farmers' perceptions of pest problems and IPM concepts and whether these could become major stumbling blocks to IPM implementation.

METHODS

The basic premise is that appropriate technological changes for small farmers must emerge from agro-socioeconomic studies that identify conditions influencing traditional cropping systems. A proper analysis must incorporate the farmers' criteria, including the way they perceive pest problems and how they react to risk.

Survey sites were selected to include cropping patterns with similar performance and shared agroclimatic characteristics: Kuala Langat (KL) in Selangor, and Beranang (BR) and Labu-Nilai (LN) both in Negri Sembilan. The three districts have been adopted by the Universiti Pertanian Malaysia as field laboratories for students to carry out their extension practicals. The farmers in these villages are therefore more exposed to modern agricultural technologies than in neighboring sub-districts.

The number of farmers interviewed was: 33 (KL), 25 (BR) and 20 (LN). The survey studied the farm system, the major ecological and technical relationships and pest management decision-making tactics of small farmers. It investigated farmers' perceptions of pests and the damage caused; control measures known and/or adopted and their effectiveness; perceptions of the environmental hazards of pesticides and their willingness to accept new pest control innovations such as IPM.

The survey was conducted in March 1984 and each respondent verbally answered questions from a questionnaire. The interviewer ticked matching answers.

The questionnaire covered the following topics: age; level of education; farming system; major crops planted; problems relating to crop production; type of control decision taken; source of recommendations; agencies recommending; level of success achieved; ability to diagnose pest problems; ability to differentiate different pesticides and their mode of action; awareness of hazards of using pesticides; knowledge of non-chemical methods of control; and perception of IPM.

Data were tabulated, classified according to survey site, and analysed using a computer.

RESULTS

Table 17.1 shows the characteristics and conditions of typical traditional small farmers in the areas studied. The low percentage of full-time farmers, in an agriculturally based country like Malaysia, occurs because farming is not usually sufficiently profitable to support an average family of 5-6 people. These farmers sought additional employment outside the farm. As the number of part-time farmers increases, this is likely to affect pest control strategies, as part-time farmers have been shown to depend heavily on prophylactic pest control (Morita, 1982). The practice of mixed cropping provides some flexibility and stability of income. The farmers perceived pest problems to be worse than fertilizer and water problems; 84% indicated that, of the various components of modern agriculture, pest management presented by far the most difficult challenge to traditional small-scale farmers as they make the transition to scientific farming. This maybe because (IPM) requires them to grasp a complex set of recommendations and data (Goodell, 1984).

TABLE 17.1

Farm and farmer characteristics for the sub-districts Kuala Langat (KL), Beranang (BR) and Labu-Nilai (LN)

Characteristics	KL	BR	LN	Overall
Mean age (years)	46	54	52	50
Primary schooling (%)	85	85	85	85
Full-time farmers (%)	52	76	68	64
Main crops	oil palm	fruits	fruits	—
	fruits	rubber	rubber	
	coffee	vegetables	vegetables	
Mixed cropping systems (%)	70	67	—	70

The farmers' concern over pest problems is shown by the high percentage (80%) taking action to control pests. Those who applied control measures were very satisfied with the results. Only 51% used chemical pesticides and most of the insecticides and fungicides were used on vegetables (Ooi et al., 1983). The percentage of small rice farmers using pesticide in the Muda Irrigation Scheme was 62% compared to only 20% of the farmers outside the Muda scheme (Normiya, 1982). The highest percentage of farmers using pesticides was in Kuala Langat (57%) where oil palm is the major crop. This was probably because of the greater need for herbicides in oil palm fields. The farmers were still dependent on non-chemical control methods because of their strong adherence to traditional cropping systems which incorporate inherent control mechanisms. However, being traditional does not mean they are not receptive to modern agricultural methods.

They found television, small group talks and personal visits by officers from the Department of Agriculture to be most useful in providing advice on pest control. The services provided by the extension department of the University Pertanian Malaysia (UPM) ranked second as the most favored agency.

Table 17.2 indicates that most farmers felt pest control was difficult but that they could diagnose pest problems. However, only 41% were actually able to differentiate between insect and non-insect attack. They were also very poorly informed about different pesticides and their mode of action. The farmers were aware of the hazards of pesticides to the environment and their health. The majority (85%) perceived that pesticides were dangerous to their health, but those who were more dependent on pesticides showed the least concern (63% from Kuala Langat, compared to 93% from Labu-Nilai and 100% from Beranang). The farmers were also well aware of the problems of phytotoxicity (73%), pesticide resistance (51%), destruction of natural enemies (50%) and chemical waste (46%).

Table 17.3 shows that half the farmers had heard of IPM, but they had a poor understanding of what it meant. In spite of this they were ready to accept IPM. Their knowledge of non-chemical pest control methods is important for the adoption of IPM and half of them were still using these methods, in the following descending order of familiarity: cultural, mechanical, physical and biological (Table 17.4). Cultural and mechanical methods are probably preferred because they use simple procedures and are readily available.

Table 17.5 indicates that farmers who were better at diagnosing pest problems or differentiating between pesticides had more favorable perceptions of IPM and other non-chemical methods, as an alternative to pest control by chemical pesticides only.

TABLE 17.2

Farmers' perceptions of pests and pesticides (% responding positively)

	KL	BR	LN
Pest control is difficult	94	68	89
Can diagnose problems	79	68	84
Know different pesticides	33	44	47
Know mode of action	7	20	27
Pesticides alone not enough	52	56	48
Should stop pesticide usage	63	33	32

Middle aged farmers (40-60 years old) showed greatest awareness of pest problems, took action to control them, referred to publications on

TABLE 17.3
Farmers' perceptions of IPM and its associated concepts (% responding positively)

	KL	BR	LN
Heard of IPM	28	47	53
Understand IPM	16	28	11
Willing to accept IPM	60	75	63
IPM is more complex	60	54	83
Ready to change the present system	67	80	55

TABLE 17.4
Farmers' use of non-chemical control methods (% responding positively)

Type of control	KL	BR	LN
Cultural	81	88	84
Mechanical	66	95	72
Physical	41	61	53
Biological	7	17	31

control recommendations, showed greater ability to diagnose problems and agreed that pesticide use should be minimized ($P < 0.05$ Chi-square test). Similarly, farmers who had formal education (at least at primary school) were more aware of pests than those who had no primary education.

DISCUSSION

In general, the small farmers interviewed in this study showed a willingness to accept new systems of pest control such as IPM and their perception of IPM as an alternative to chemical control, or any other unilateral approach to pest problems, was favorable. The potential of cultural and mechanical methods of pest control, as a major component of IPM on small farms should not be ignored. With increased knowledge of ecological and biological methods of control, through education and the participation of middle-aged farmers, the available labour and other resources should be adequate for the implementation of IPM in the survey areas.

TABLE 17.5

Association between farmers' understanding of pest control and their perceptions of non-chemical control methods†

	Cultural	Mechanical	Physical	Biological	Integrated
Can diagnose problems	S	S	S	S	S
Can differentiate pesticides and mode of action	S	NS	NS	S	NS
Problem with resistance	S	NS	NS	S	S
Cost of pesticide	—	—	—	—	S

† Chi-square test: S denotes significance at $P = 0.05$ and NS no significant association.

The fact that farmers have a favorable attitude towards IPM, even although they have little understanding of what it means may be because they perceive that IPM does not impose an extra burden on their limited resources. They may view it like a bottle of pesticide which will be made available and require little of the farmer's own control and little extension, whereas in practice IPM is actually more labour intensive and requires good extension and more scientists (Goodell, 1984). The small farmers are becoming too dependent on technological packages, like pesticides, and thus have little experience and skill in methods of managing their own resources which may be laborious or difficult to learn, such as cultural control.

Pest perceptions among the farmers in this survey may differ from those of other areas in Malaysia. Vegetable farmers in the Cameron Highlands, for example, Pahang, viewed pests more seriously than rice farmers. They are probably more risk averse and used insecticides on a fixed schedule, whereas rice farmers in Tanjung Karang, Selangor (Heong, 1982) and in the Muda Irrigation Scheme, Kedah, applied insecticides only after pest damage was visible (Heong et al., 1983).

The farmers in this study seemed to be receptive to new technology which would allow them to make their own decisions, provided it did not impose additional costs.

In studying farmers' resistance to IPM, Sheahan (1980) adopted the systems approach in which IPM is perceived as one sub-system of the production unit, the farm. Within this sub-system, there are further sub-systems for each plant pest. IPM is often perceived as the most complex sub-system and is not fully understood by farmers, extension workers or even research staff. The other sub-systems which could influence farmers' rejection or acceptance of IPM are the social sub-system, the economic sub-system, the physical resources sub-system, and the cultural practices sub-system.

Unless farmers perceptions are understood, and shown to have practical value in making pest management decisions, implementation of IPM programs will be met with unexpected obstacles. Researchers should therefore take small farm technology back to the drawing board, to incorporate traditional technologies (Norton, 1982b; Hussein, 1983; Norton and Mumford, 1983; Goodell, 1984; Matteson et al., 1984).

The government has tended to provide agricultural packages and bureaucratically generalized services incompatible with the principles of IPM, on the assumption that farmers cannot make rational decisions on their own behalf. If farmers are not trained to make their own decisions, they cannot adopt IPM. The pest surveillance system for brown plant hopper management on rice in the Tanjung Karang irrigation scheme in Malaysia has violated a basic IPM principle. All surveillance data collected from farmers' fields are transported to a computer, 100 km away in the capital city of Kuala Lumpur for compilation, pre-empting the initiative and responsibility of the field technicians and farmers. IPM is incompatible with this, and many other forms of state control and agricultural subsidy in developing countries.

REFERENCES

Goodell, G. (1984). Challenges to international pest management research and extension in the Third World: Do we really want IPM to work? *Bulletin of the Entomological Society of America* 30(3): 18-26.

Heong, K.L. (1982). Pest control practices of rice farmers in Tanjung Karang, Malaysia. *Insect Science and its Application* 5(3): 221-226.

Heong, K.L., Ho, N.K. and Jegathessan S. (1983). The perceptions and management of pests among rice farmers in the Muda Irrigation Scheme, Malaysia. (Unpublished paper).

Hussein, M.Y. (1983). Integrated pest management and the small farmers of Malaysia. International Conference on Hazards of Agrochemicals in the Developing Countries, 8-12 Nov. Igsa, Alexandria, Egypt.

Matteson, P.C., Altieri M.A. and Gagne W.C. (1984). Modification of small farmer practices for better pest management. *Annual Review of Entomology* 29(1): 383-402.

Morita, K. (1982). *The Disease and Insect Pest Occurrence Forecast Program in Japan*. Report of the 11th Session of FAO/UNEP Panel of Experts on Integrated Pest Control, Kuala Lumpur' March 1982.

Normiya, R. (1982). Problems in transfer, delivery and acceptance of rice technology. *MARDI Rural Sociology Bulletin* 12, (In Bahasa Malaysia).

Norton, G.A. (1982a). Crop protection decision making — an overview. In *Decision Making in the Practice of Crop Protection,* Monograph No. 25, ed. E.B. Austin, Croydon, London: BCPC Publications. pp.3-11

Norton, G.A. (1982b). A decision-analysis approach to integrated pest control. *Crop Protection* 1: 147-164.

Norton, G.A. and Mumford, J.D. (1983). Decision making in pest control. In *Advances in Applied Biology,* VIII ed. T.H. Coaker, 57:113, London: Academic Press.

Ooi, A.C.P., Heong, K.L., Lim. B.K. and Mazlan, S. (1983). Adoption of pesticide application technology by small-scale farmers in Peninsular Malaysia. In *Pesticide Application Technology,* eds. G.S. Lim and S. Ramasamy, 148-158, Kuala Lumpur: MAPPS.

Sheahan, B.T. (1980). *Producer Resistance to Integrated Pest Management: An Extension View-point.* 50th ANZAAS Congress, May 1980, Adelaide, South Australia.

18

Pest Control Practices And Pesticide Perceptions Of Vegetable Farmers In Loo Valley, Benguet, Philippines

Charito P. Medina

INTRODUCTION

Benguet province, in the Cordillera Mountains in northern Philippines, is very rugged with elevations ranging from 1,600 to over 2,300 m. The high altitude results in a mean annual temperature of 18°C, 9°C lower than the country as a whole, making the area highly favorable for vegetable growing, particularly semi-temperate and temperate vegetables.

Buguias is the main vegetable producing area of Benguet, with 35% of the total land area planted to vegetables and 87% of the total labor force engaged in agriculture (Benguet Socio-economic Profile, 1981). The most extensive vegetable terraces are found in the Loo Valley, where the survey data reported in this paper were collected. However, the practices and perceptions described are typical of vegetable farmers throughout the Cordillera.

The vegetable industry which is both labor and capital intensive, was introduced to the Loo Valley in the late 1950s by immigrant Chinese merchants and stimulated by market demand. Mountain sides were cleared and terraced to accommodate vegetable growing. Only one crop per year is grown in the sloping rainfed areas, but on the irrigated terraces four crops are possible. The result is that crops at every stage of growth can be found in the field at any given time.

This intensive farming system supported the build up of pest populations by providing unlimited food, and the lack of spatial and temporal barriers favored the maintenance of high population levels. Thus pests and diseases are very severe throughout the year.

PEST CONTROL PRACTICES

Some farmers recalled that they used to remove 'worms' from plants by hand-picking. They also reported using plant decoctions to control insect pests, e.g. avocado leaves for ants, sunflower for diamond-back moth, and also tobacco and hot pepper decoctions. With the exception of detergents, which are sometimes sprayed to control leeches, such methods are rarely practised now.

Chemical pesticides are now the only means of crop protection used by farmers, having gained their confidence because of their immediate and dramatic effect. They are readily available in agricultural supply stores and are necessary to maintain the cosmetic quality standards demanded by consumers, who have been influenced by chemical company advertisements.

SOURCES OF INFORMATION ON PESTICIDES

Information on pesticides comes mainly from chemical company fieldmen who organize 'Farmers' Vegetable Seminars' through the extension personnel of the Ministry of Agriculture and the head of the local administrative unit (the barangay captain). These seminars are used to promote product sales. The format is a short academic discussion on the kinds and biology of pests, followed by information on relevant pesticides produced by the company. At the end of the seminar, pesticides are sold at a special promotion price, often with added incentives like raffles for T-shirts bearing the names of pesticide products, small packages of pesticides, tumblers or flashlights distributed free.

To a more limited extent, information also comes from other farmers, billboards or radio advertisements, demonstration methods or from the government extension agents (Bahatan et al., 1970).

PESTICIDES USED

Farmers tend to spray the same very wide range of pesticides on all their crops. Chemicals are usually sprayed in combination and the efficacy of one may mask the inefficacy of others in the mixture. Farmers believe that such mixtures are good, without questioning whether some chemicals may not be needed in the mixture.

On six vegetable farms, 90% of pesticide treatments on potatoes were applied as mixtures of insecticides and fungicides and a similar pattern was found on chinese cabbage (Table 18.1). On lettuce, sweet peas and beans, a combination of insecticides and fungicides was also used. When biological insecticides were used, they were generally combined with other insecticides because, according to farmers, "Biological insecticides cannot kill the worms inside tunnels or those inside leaf whorls, and the worms are not dead the following day". The farmers did not understand that caterpillars have to ingest the microbial spores to be killed hence the delay. The pesticides used by farmers are listed in Table 18.2.

TABLE 18.1

Pesticide combinations per spray round, from records of six farmers, 1983

Crop	Pesticide Combinations	Percent of total sprays
Potato	(Total no. of pesticide treatments = 52)	
	2 insecticides + 1 fungicide	46
	1 insecticide + 1 fungicide	44
	1 fungicide	10
	(23% included urea)	
Chinese cabbage	(Total no. of pesticide treatments = 56)	
	2 insecticides + 1 fungicide	46
	1 insecticide + 1 fungicide	27
	2 insecticides	23
	3 insecticides + 1 fungicide	2
	3 insecticides	2
	(28% included urea)	

Mixing of pesticides is encouraged by the farmers' desire to have rapid knockdown of pests. It is also recommended by chemical company salesmen as one way of preventing pest resistance. This idea is questionable, at least as practised, because the combinations used are indiscriminate. It does, however, increase pesticide sales. Farmers usually use 1-4 tablespoonfuls of each chemical per load in a 19 litre (5 gal) sprayer so that with mixtures the amount per sprayer load totals 3-8 tablespoons. The strength and volume of the mixture applied is increased as the crop matures, to compensate for the increasing size of plants.

In a growing season of 65 to 95 days, farmers spray an average of eight to ten times with three to six kinds of pesticides on a calendar basis. The usual spraying interval is seven days, but this is reduced to three days during summer for pests like thrips, or during the rainy season for blight diseases. The spraying records of representative farmers are summarized in Table 18.3.

PERSONAL PROTECTION AND PESTICIDE POISONING

Spraying is usually done in the early morning, or late afternoon. Farmers use very little personal protection during spraying, the maximum being boots and a piece of cloth covering the mouth. Some wear short sleeved shirts and no gloves, and barefooted farmers even fold up the lower portion of their pants so that they do not get wet. As a consequence, their legs and feet, hands come into contact with pesticides. Some farmers even use their bare hands to mix pesticides in a container, claiming that this is more effective.

TABLE 18.2
Pesticides applied to vegetable crops (1983)

Crop	Insecticides	Fungicides
Potatoes	*Bacillus thuringiensis* cypermethrin fenvalerate formetanate malathion methamidophos methiocarb mevinphos oxamyl profenofos triazophos phenthoate	chlorothalonil cymoxanil mancozeb maneb metalaxyl zineb
Chinese cabbage	*Bacillus thuringiensis* cartap cypermethrin endosulfan endrin fenvalerate malathion methamidophos mevinphos propoxur triazophos phenthoate	cymoxanil mancozeb maneb
Sweet peas, Banguio, beans, lettuce	*Bacillus thuringiensis* cartap fenvalerate formetanate propoxur triazophos phenthoate	cymoxanil mancozeb maneb

Wind direction is not considered during spraying. Farmers spray along rows, back and forth, even with the wind. Wind velocity is also not considered and it is not uncommon to smell pesticides drifting up to 200m away from a spraying farmer.

After spraying, farmers wash their sprayers near or in irrigation canals. They also wash their hands and feet in the same canals, without

TABLE 18.3
Spraying intervals for potato and chinese cabbage (records from six farmers, 1983)

Case	Cropping months	Spraying intervals (days)†	No. of sprays	Growing period (days)
Potato				
1	Feb-May	18-6-7-5-10-8-7-34	7	95
2	Jan-Apr	33-5-6-5-5-6-7-26	7	93
3	Apr-Aug	17-4-5-5-5-5-7-4-3-39	9	94
4	Feb-May	24-11-12-3-7-7-8-17	7	89
5	Jan-Apr	14-8-7-7-6-7-7-7-7-7-7-7-27	12	104
6	Aug-Nov	11-6-9-4-3-8-7-3-5-22	9	78
7	Dec-Mar	26-6-6-10-4-8-20-15	7	95
8	Oct-Dec	20-3-3-3-5-3-7-4-3-6-27	10	84
Chinese Cabbage				
1	May-Jly	11-3-9-5-3-10-17-15-12	8	73-85
2	Feb-May	20-11-15-7-7-4-7-15-8	7-8	86-94
3	May-Jly	7-3-5-4-5-5-4-5-7-4-18	10	70-77
4	Feb-May	24-11-12-3-7-7-8-17	7	89
5	Sept-Nov	6-6-8-4-11-7-3-5-17-16	9	83

† First number is interval from planting to first spraying; last number is interval from last spray to harvest.

detergent or soap. These irrigation canals are also used for washing clothes or bathing children downstream. Bathing after spraying is seldom practised. Farmers who become dizzy due to inhalation of pesticides just sleep until they recover.

Re-entry regulations are not observed after spraying. In some cases, the houses are built in the middle of the vegetable terraces and the farms are also the children's playground.

Empty pesticide containers are collected and dumped, uncovered, in backyards. Empty pesticide boxes and plastic bags are often left in the field and carried by wind until they settle in canals.

Records of pesticide poisoning from the Lutheran Hospital revealed that there had been 80 cases from the Loo Valley between January 1967 and June 1984, an average of 4.4 people poisoned annually. The 1983 census recorded a total population of 2,060 and therefore about 0.2% of the total population is poisoned each year. The cumulative percentage of poisoned people, up to June 1984, is therefore 3.9% of the total population.

The age of victims ranged from 10 months to 48 years with 49% between the ages of 11 and 20, 30% between 21 and 30 years old, and the remaining 21% either below 10 or above 30 years old. The sex distribution of victims was fairly equally balanced, 59% males and 41% females. Recovery time ranged from 1-4 days, with most (54%) having only one day's confinement. Three of the 80 cases died. Poisoning occurred by ingestion (28%), inhalation (30%), suicide attempts (13%), and the rest were unspecified. All suicide cases recovered.

In interviews with family heads, about 61% reported that they had experienced pesticide poisoning of some kind (Table 18.4). Of these, many had medical attention (Table 18.5), but most practised self-treatment or just slept. Self-treatment included taking medicine to relieve the immediate symptoms, such as a pain reliever for headache, medicine to relieve diarrhoea or smoking a cigarette for vomiting. Others took herbal medicine, e.g. tea, while a few reported drinking dry gin. The latter practice could explain why several poisoning records reported alcoholic breath.

In a case study of 27 farmers, the symptoms experienced were itchy skin, rashes, dizzy spells, vomiting or nausea, headache, eye irritation and loose bowel movement. However, most never went to the doctor for a consultation or check-up. A doctor was only consulted if the poisoning was very serious. Otherwise they tended to sleep for a few hours until they felt better. These results imply that the officially recorded estimates of the numbers of poisoning cases discussed above are very low, if the number of unreported cases is taken into account. Many cases of sub-lethal, secondary and cumulative poisoning are not included in these statistics.

FARMERS' PESTICIDE PERCEPTIONS

The vegetable farmers are more familiar with the names of pesticides than with the pests they are controlling. Pesticide spraying provides a feeling of security to farmers who have invested a lot in seedlings and fertilizer. Pesticides are their only perceived pest control measure and when they ask for different control measures, they are referring to other brands of chemical. New brands are usually regarded as promising and they are willing to try them.

Different brand names are also regarded as different chemicals. One farmer mixed mevinphos with Backie and Dipel, two different brands of *Bacillus thuringiensis* spores. Similar combinations are common in the area.

The decline in the efficacy of pesticides that have been used before is usually blamed on chemical companies. However, this could be due to the development of resistant pests or a decline in toxicity of the chemicals due to poor storage by the farmers.

Pesticide poisoning is not considered important by the farmers unless it is serious enough to warrant hospital treatment. They either do not give attention to pesticide residues and pesticide accumulation in the environment and their own bodies or they may not be aware of them.

TABLE 18.4

Number reporting pesticide and pesticide-related symptoms in Loo, Buguias †

	With symptoms	Without symptoms
Headache	252	12
Vomiting	163	69
Dizziness	162	79
Diarrhoea	203	51
Skin disease	182	69
Otitis media ‡	90	146
Itchy eyes	178	65
Direct pesticide poisoning	150	94

† Total sample of household heads in the study site(363); interviews done by Cordillera Studies Center in May-July, 1984 (many interviewees experienced more than one of the symptoms quoted).
‡ Disease of the ears involving a secretion of mucus.

TABLE 18.5

Number of respondents who experienced the symptoms in Table 18.4 taking specific actions

Symptoms	Action Taken			
	Medical consultation	Self-treatment	Rest/no medicine	Others
Headache	76	144	21	11
Vomiting	72	42	25	8
Dizziness	62	40	56	4
Diarrhoea	76	119	4	4
Skin disease	85	79	2	16
Otitis media	64	22	1	3
Itchy eyes	86	89	1	2
Direct pesticide poisoning	20	48	76	6

CONCLUSION

There is much overuse, misuse and abuse of pesticides by vegetable farmers in the Philippine Cordillera. Spraying pesticides has now become a habit, rather than a necessity, and spraying is done indiscriminately. Under these circumstances, pesticide usage in the future is bound to increase further, with a corresponding increase in cases of pesticide poisoning.

ACKNOWLEDGMENTS

My thanks to the research staff from the Program on Environmental Science and Management (PESAM), University of the Philippines at Los Banos (UPLB) and Cordillera Studies Center (CSC), U.P. Baguio who composed a multidisciplinary team. This paper is part of a research on vegetable growing in the Loo Watershed Area conducted by the team. Support was provided by the Ford Foundation.

REFERENCES

Bahatan, F.D. Jr., Bahingawan, H., Cosalan, P.M. and Tejano, A.R. (1970). An analysis of the programming behavior of vegetable farmers in the province of Benguet. *Saint Louis University Research Journal* 1(3) : 503-580.

Benguet Socioeconomic Profile (1981). Baguio Printing and Pub. Co., Inc., Baguio City, 280pp.

19

Towards The Management Of Pests In Small Farmer Mixed Cropping Situations In Tropical Africa

Anthony Youdeowei

INTRODUCTION

Throughout tropical Africa, small-scale farmers continue to use traditional methods to cultivate their meagre land resources. In Nigeria these farmers constituted 18% and 15% of the total population in 1965 to 1984 and the projected figure for 1995 is 12%. Peasant farmers produce at least 80% of the food consumed in West Africa, and efforts to increase national food production must involve inputs and programs directed at helping them. This point was emphasised by Trew (1978) as follows:

> After fifteen years of international aid, it has become obvious to me that the major hindrance to rural development is the lack of a policy aimed specifically at helping the small farmer evolve. To be sure, there are innumerable rural development projects being carried out, but very few have been planned specifically to serve the immediate needs of native farmers. Even fewer take full account of the limitations resulting from his illiteracy, his limited means, and especially the conflict of traditions and politics with development.

Early efforts to increase food production concentrated on developing technology for large-scale farming, often omitting consideration of the social and economic systems in rural communities. It was erroneously believed that traditional multiple cropping systems were less productive than monoculture, were out-dated and were only a transitional stage in progress towards large-scale monoculture. However, it is now obvious that multiple cropping is a vital factor in the food supply system in tropical Africa and that agricultural production by this means can be greatly increased. This paper describes some initial results of research at Ibadan, Nigeria, on insect pests on small farms with mixed cropping, and discusses an approach towards development of a pest management strategy relevant to such systems in tropical Africa.

SMALL FARMS IN TROPICAL AFRICA

The small farm in tropical Africa is a complex ecosystem, varying in size from 0.001 ha to about 0.45 ha. Food crops such as maize, cassava, yams, potatoes, sorghum, cowpea, okra, melon, tomatoes and a variety of leafy vegetables, along with some medicinal crops, are cultivated in traditional mixed inter-cropping systems combined with shifting cultivation. Surpluses are sold for cash to buy goods and commodities which cannot be produced on the farm, to pay for the education of children and relatives, or to meet traditional and social needs. Farmers may also keep livestock such as sheep, goats, pigs and poultry for meat sales, manure and other purposes.

The system thus provides the farmer with a variety of produce, optimizing the use of limited land resources and providing an insurance of a reasonable harvest in at least one food crop. Tools are simple, requiring mainly human labour, the financial outlay is small and farm inputs such as fertilizers and pesticides are used on a limited scale. However, yields are usually much lower than the potential for the crop. Mixed cropping is reputed to discourage the build up of pest populations, but economic damage to crops often occurs, requiring the use of chemicals to control pests. Research at the International Rice Research Institute (IRRI) in south-east Asia has shown that when pesticides are used in mixed cropping situations, the level of pest control achieved was much lower than in monocultures (IRRI, 1973). In mixed cropping systems, extra care is necessary to ensure that, while destroying the target pests, beneficial insects are not adversely affected. Also, in mixed cropping systems, a pesticide recommended for a particular crop may be phytotoxic to other crops.

Small farmers may not have the money to pay for pesticides, in the absence of a subsidy scheme. Even more serious is the lack of information on pesticide application and on the introduction of integrated pest management (IPM) into mixed cropping systems. Research is needed on the development of integrated pest control in a farm suitable for small farmers in Africa. This paper will consider some relevant pest problems and suggest an approach to the development of an appropriate pest management system.

OBJECTIVES AND APPROACH

In 1981, in Ibadan, Nigeria, research was begun with the following objectives: a) to understand the cropping patterns and agronomic practices of small farmers and the influences of crop production factors on yields; b) to understand the diversity of the insect populations in the mixed cropping systems and to establish the pest status of various insects; c) to examine pest damage on specific crops and estimate economic thresholds for individual pests; d) to analyse the cropping patterns and agronomic practises in relation to their influence on insect pests, crop damage and yield. Integrated pest management strategies which take account of interactions between cropping pattern, insect pest density and behavior would then be developed to provide a long term, environmentally acceptable solution to the pest problems of small farmers.

PEST PROBLEMS ON SMALL FARMS

Four mixed cropping farms on the University of Ibadan campus were selected for study. These were cultivated by small-scale farmers to meet the food requirements of their households and to generate extra income. The farms had produced a variety of crops each growing season for three consecutive years, including root and tuber crops, vegetables and to a lesser extent cereals and grain legumes.

Surveys indicated that a wide range of insects attacked the crops throughout the year, as illustrated in Figure 19.1 (Udosen, 1982), with varying degrees of damage. On one farm the egg plant *Solanum melongena*, was the most heavily attacked with 112 insect pests per plant. The nature of pest damage included leaf skeletonization due to *Hymenia recurvalis*, defoliation of cassava *Manihot esculenta* by *Zonocerus variegatus*, and sucking damage to the inflorescence and young seeds of the leafy vegetable *Amaranthus hybridus* by pests such as *Cletus fuscescens*. In the latter case, *C. fuscescens* attacked only the developing seeds, which subsequently became shrivelled, flat, and light brown in color, compared to the dark purple color of undamaged seeds. An infestation of 17 insects per plant resulted in a 50% reduction in seed germination, adversely affecting the source material for the following year's plantings (Olatunji, 1982). Damage by *Podagrica* to okra plants was significantly higher (79%) when they were in a block rather than sparsely distributed (43%) amongst other crops in the farms. Similarly, where *A. hybridus* was planted in a block, leaf damage by *Gasterodisus rhomboidalis* (Curculionidae) was 53%, while in another plot where it was dispersed among other plants, the average damage level was 33% (Figure 19.2) (Peter-Paul, 1983). Thus, the intensity of insect pest damage depends to some extent on the type and pattern of cultivation of the host crop.

Further research will focus on the effect of cropping pattern on minimizing levels of pest damage.

THE DEVELOPMENT OF IPM FOR SMALL FARMS

Although IPM strategies have been developed for a variety of farming situations, little seems to have been done to introduce it to small farmers in tropical Africa. Where an IPM strategy has been developed for some key pests, the essential political, socio-economic and environmental considerations have been neglected, limiting the success of the programs. Recent evidence indicates that small farmers in Africa will increasingly depend on pesticides alone to control insect pests without a satisfactory understanding of the associated hazards (Youdeowei and Service, 1983). IPM could enable them to produce higher crop yields more cheaply, and maintain the quality

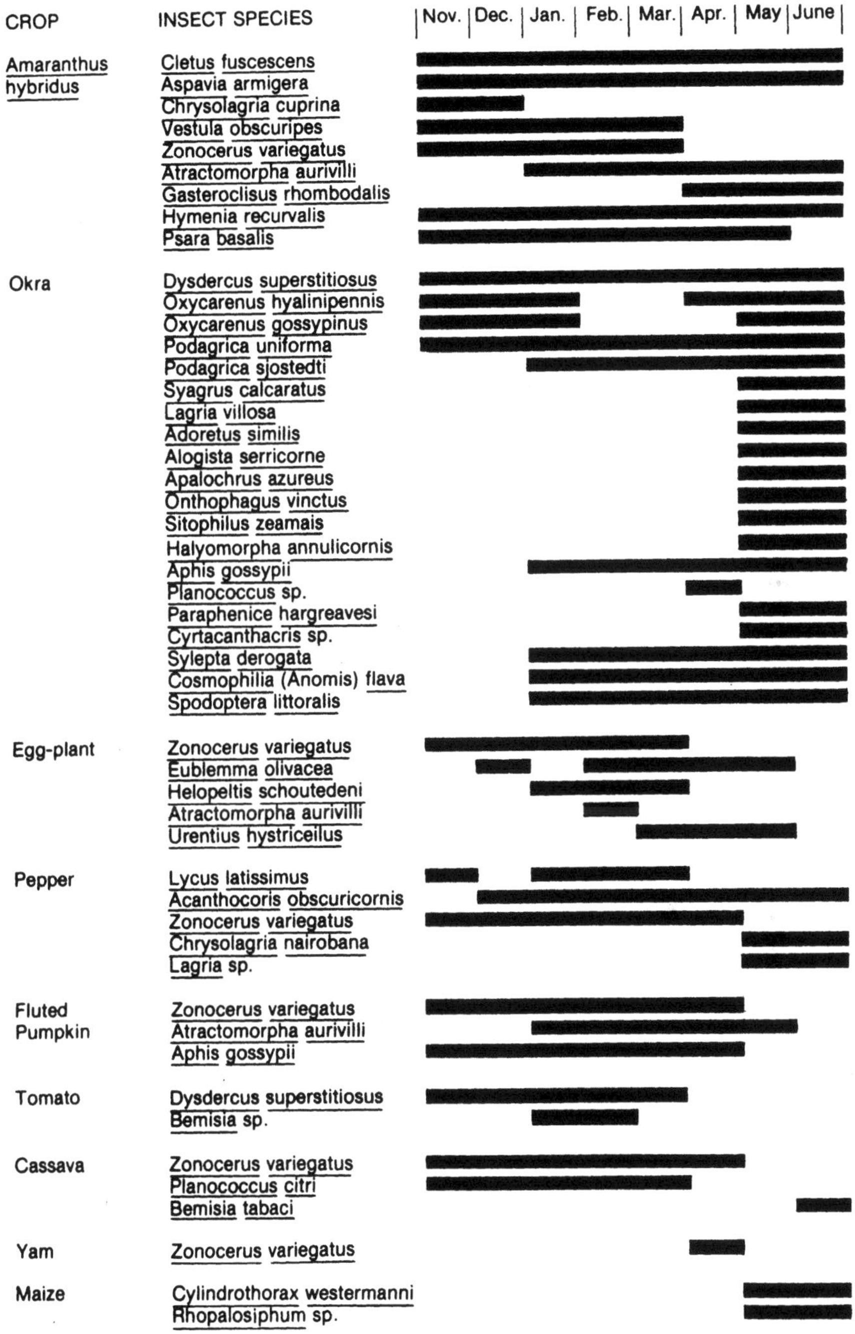

CROP	INSECT SPECIES	Nov.	Dec.	Jan.	Feb.	Mar.	Apr.	May	June
Amaranthus hybridus	Cletus fuscescens								
	Aspavia armigera								
	Chrysolagria cuprina								
	Vestula obscuripes								
	Zonocerus variegatus								
	Atractomorpha aurivilli								
	Gasteroclisus rhombodalis								
	Hymenia recurvalis								
	Psara basalis								
Okra	Dysdercus superstitiosus								
	Oxycarenus hyalinipennis								
	Oxycarenus gossypinus								
	Podagrica uniforma								
	Podagrica sjostedti								
	Syagrus calcaratus								
	Lagria villosa								
	Adoretus similis								
	Alogista serricorne								
	Apalochrus azureus								
	Onthophagus vinctus								
	Sitophilus zeamais								
	Halyomorpha annulicornis								
	Aphis gossypii								
	Planococcus sp.								
	Paraphenice hargreavesi								
	Cyrtacanthacris sp.								
	Sylepta derogata								
	Cosmophilia (Anomis) flava								
	Spodoptera littoralis								
Egg-plant	Zonocerus variegatus								
	Eublemma olivacea								
	Helopeltis schoutedeni								
	Atractomorpha aurivilli								
	Urentius hystriceilus								
Pepper	Lycus latissimus								
	Acanthocoris obscuricornis								
	Zonocerus variegatus								
	Chrysolagria nairobana								
	Lagria sp.								
Fluted Pumpkin	Zonocerus variegatus								
	Atractomorpha aurivilli								
	Aphis gossypii								
Tomato	Dysdercus superstitiosus								
	Bemisia sp.								
Cassava	Zonocerus variegatus								
	Planococcus citri								
	Bemisia tabaci								
Yam	Zonocerus variegatus								
Maize	Cylindrothorax westermanni								
	Rhopalosiphum sp.								

Figure 19.1   Insect species attacking crops on a farm in Ibadan, Nigeria

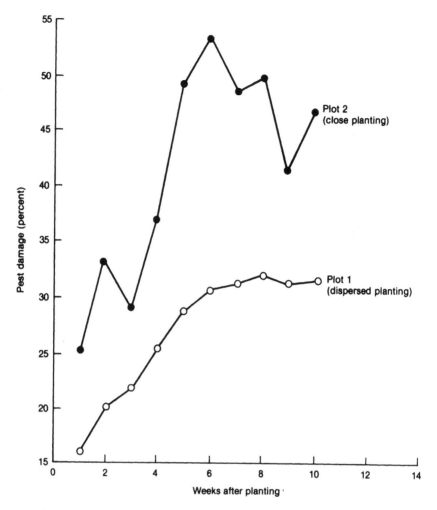

Figure 19.2   Percent of leaves of <u>Amaranthus</u> <u>hybridus</u> damaged by <u>Gasteroclissus</u> spp. in small farmer mixed cropping farms in Nigeria (Peter-Paul, 1983)

of their environment.

In considering IPM for small farmers two interacting systems are envisaged, the crop production system, and the pest damage system.

Interdependent components of the crop production system include crop phenology, cropping pattern, crop requirements (in terms of soil moisture and nutrients, precipitation, photo-period and insolation), fertilizer and pesticide application, and irrigation.

In the pest damage system accurate identification of the pests is crucial for any control strategy. Other components include: recognition of damage symptoms; understanding of damage thresholds; identification of natural enemies, including parasites, predators and pathogens; and knowledge of sampling techniques for monitoring pest populations.

IPM is based on knowledge of the interactions between the crop production and pest damage systems.

It is thus important to know how the farmer prepares his land to grow his crops, and why he grows his crops in a particular way in his locality. To what extent do his farming practices alter the environment and create pest problems? When do pests arrive and how do they damage the crops? Is the farmer aware of the influence of his agronomic practices on the dynamics of pest populations? What is the farmer's perception of pests and their control? Which methods of pest control does he practice? What natural enemies are present in the mixed cropping ecosystem? How do the planting system and the agronomic practices affect the populations and the presence and importance of natural enemies?

There is now an urgent need for data on the crop production and pest damage systems of small farmers in Africa to be organized into a form suitable for the formulation of low technology, cheap and efficient IPM protocols. IPM systems should be tested on small farms and modified if necessary. The introduction of IPM will require training in the concepts and methods of IPM on the farms themselves and organization of the farmers into IPM groups. Training should be done by both research scientists and extension specialists, with emphasis on group action by the farmers and successful trainees should subsequently be recruited as trainers. This will generate closer contacts and improved confidence between research workers and the farmers. National governments, through the Ministries of Agriculture and Rural Development, in collaboration with universities and national or international research institutes should take responsibility for research and training on IPM for small farmers in tropical Africa.

ACKNOWLEDGMENTS

I would like to acknowledge the travel grant enabling me to attend the fourth International PMPP Conference in Thailand, which was presented by Clark University, using funding from the General Services Foundation, USA.

REFERENCES

International Rice Research Institute (IRRI), (1973). Cropping Systems Program. Annual Report 1973. IRRI, Los Banos, Philippines.

Olatunji, O.G. (1982). Damage assessment of *Cletus fuscescens* Walk. (Heteroptera, Coreidae) in relation to *Amaranthus overaceus*. M.Sc. Thesis, University of Ibadan.

Peter-Paul, S. (1983). Assessment of insect pest damage to some vegetables in tropical small farmer mixed cropping systems. M.Sc. Thesis, University of Ibadan.

Trew, M. (1978). The small farmer: A neglected resource. *Society for International Development Focus* 2: 13-17.

Udosen, T.E.G. (1982). Preliminary studies on the development of pest management systems in small farmer mixed cropping farms in Nigeria. M.Sc. Thesis, University of Ibadan.

Youdeowei, A. and Service M.W. (1983). *Pest and Vector Management in the Tropics*. Longman Group.

20

Farmers' Perceptions Of The Rice Tungro Virus Problem In The Muda Irrigation Scheme, Malaysia

K.L. Heong and N.K. Ho

INTRODUCTION

Successful implementation of pest control requires adequate knowledge of pest biology, ecology and control measures and also, information on how the pest is perceived by farmers, their attitudes towards pest problems, their beliefs and the control measures adopted. Appropriate socio-economic research is an important component of integrated pest management (IPM) (e.g. Tait, 1978a; b; Litsinger et al., 1980; Prasadja and Ruhendi, 1980; Mumford, 1981; Heong, 1984; Heong et al., 1985), providing better understanding of farmers' decision making patterns and helping in the deployment of scarce research expertise and resources. In fact, these surveys should precede applied research to ensure it is not merely of academic interest (Norton, 1982; Mumford and Norton, 1984; Matteson et al., 1984).

Tungro (called *penyakit merah* virus, PMV or 'red disease' in Malaysia) is a virus disease of rice transmitted by the green leafhopper (GLH), *Nephotettix virescens*. The symptoms were first reported in the 1940s and initially it was thought to be caused by physiological deficiencies (Thompson, 1940; Johnston, 1954). In 1964 the cause was demonstrated to be a virus (Ou et al., 1965), and further evidence that the *penyakit merah* virus in Malaysia and tungro found in other countries are the same disease was provided by Lim (1969) and Ting and Paramsothy (1970). Subsequent research has concentrated on virus characterization (Ling, 1972), disease epidemiology (Lim et al., 1974; Ling et al., 1982), virus host relationships (Ting and Paramsothy, 1970; Ling et al., 1983) and the ecology of the host, GLH (Lim, 1969). Little work has been directed at how farmers perceive the PMV problem and their traditional beliefs in controlling it. If we wish to influence crop protection decisions among farmers, the researchers, extension officers and the commercial agents should recognize factors affecting decision making at the farm level.

This paper discusses the results of a farm survey carried out in the Muda Irrigation Scheme in August 1984, studying how rice farmers perceive the PMV problem, their responses to it and to the government's efforts in containing the spread of the disease.

The Muda Irrigation Scheme, in north western peninsular Malaysia, covers about 96,000 ha of padi land and is farmed by 63,000 families (Afiffuddin, 1977a; b; Ho, 1979). It is Malaysia's largest rice growing area, responsible for about 50% of Peninsular Malaysia's total production. A quasi-government body, the Muda Agricultural Development Authority (MADA) is responsible for administration and agricultural extension in the scheme.

In Peninsular Malaysia, the tungro virus was endemic only in the Krian district and Wellesley province. Although *penyakit merah* symptoms had been observed in the Muda, the presence of tungro virus disease was not confirmed until early 1981, initially affecting about 6,000 ha in the south. Factors contributing to the absence of tungro were low populations of GLH and the species composition. In the past more than 60% of the GLH collected from light traps in the Muda were *Nephotettix nigropictus*, a poor vector of the virus, compared to almost 100% *N. virescens* in the Krian district (Lim et al., 1974). Another important factor was the synchronous planting of rice, imposed by the tight water schedule in the Muda. As a result, the nursery periods (which favor GLH development) were kept to a minimum. In 1981-83 the planting periods in the Muda were significantly prolonged, resulting in longer nursery periods. GLH populations were not only much higher but were also predominantly *N. virescens* and these factors probably played a significant role in the spread of tungro in the Muda. In response to the epidemic, control campaigns were organized by the agricultural authorities (Table 20.1).

METHODS

Data were collected using a formal survey with a questionnaire in Bahasa Malaysia. Farm interviews were carried out by 27 technicians closely supervised by one of us (NKH). Ten farmers were randomly selected by each interviewer (total 270). Photographs of tungro infected rice plants, the vector GLH, and sample copies of the poster and extension leaflet on tungro were shown to farmers during the interview.

The data were coded and analysed using the statistical package, Statistical Analysis System (SAS, 1982), the analyses being based on the number of farmers responding to each question rather than the sample total, i.e. farmers not responding or providing insufficient replies were ignored. In addition to the formal survey, supplementary information was obtained from records maintained by MADA, farm visits and observations of the authors, particularly NKH.

TABLE 20.1
Area damaged by *penyakit merah* and activities of control campaign in the Muda Irrigation Scheme

	Years			
Activities/Items	1981	1982	1983	1984
Damage				
Area affected by the disease (ha)	5,884	5,839	8,655	501
Area with 75-100% yield loss (ha)	125	1,983	2,006	50
Estimated loss ($US million)	1.8	4.0	4.1	0.2
Campaign Activities				
Field demonstrations to farmers				
(number of occasions)	74	274	574	432
Total farmers involved	4,862	8,258	16,853	12,383
Briefings at village mosques				
(number of occasions)	27	100	100	14
Mobile vehicle broadcasting				
(number of occasions)	27	276	281	50
Mini exhibitions in villages	—	29	32	—
Destroying diseased plants ('000 ha)				
(remove & burn)	—	64.0	72.8	67.5
(spray herbicide)	1.0	2.7	12.0	34.6
Distribution of posters	—	10,000	—	
Distribution of leaflets	—	30,000	—	
Distribution of insecticides				
(carbofuran 2G) ('000 kg)	20.0	64.8	56.7	79.8
(BPMC) ec ('000 litres)	1.0	0.07	6.7	1.5

RESULTS

Farmers' Experience and Understanding of Penyakit Merah

When farmers were shown the photograph of tungro infected rice plants, about 77% said that the plants were affected by *penyakit merah*. About 10% said that the plants were attacked by insects, while only 6% did not recognize the problem (Table 20.2). When the photograph of the GLH was shown, about 85% (n=265) of the farmers recognized the insect, but less than 70% knew that it caused *penyakit merah* (10% said that the two were not related, while 20% did not know). A range of answers was reported in a further question on the cause of *penyakit merah* (Table 20.3), but none of the farmers mentioned 'virus'.

Of 233 responding farmers, 58% had experienced *penyakit merah* in the last ten years and most of these (95%) had attempted to control the

TABLE 20.2

Recognition of *penyakit merah* infected plants from a photograph in the Muda Irrigation Scheme (% farmers responding)

Answers given by farmers	Districts			
	I n=46	II n=89	III n=59	IV n=70
penyakit merah infected	70	81	88	67
Attacked by insects	11	10	2	16
Fertilizer deficiencies	2	1	2	4
Bad weather	4	0	0	0
Infected by some disease	0	0	2	3
Water contamination	0	0	0	1
Miscellaneous answers	0	3	2	3
Do not know	13	5	5	6

TABLE 20.3

Reported causes of *penyakit merah* (% farmers responding)

Causes reported by farmers	Districts			
	I n=45	II n=89	III n=52	IV n=70
Insects	44	54	67	74
Unknown factors	2	3	8	4
Some diseases	20	7	0	1
Fertilizer deficiencies	2	3	0	2
Bad weather	4	2	0	0
Contaminated water	0	2	2	1
Insecticides	0	2	0	0
Do not know	24	24	15	13

problem. However only 61% of these farmers reported that the plants recovered after treatment. The control measures used included insecticide (33%), traditional methods (33%), burning (11%), initiating control early (11%), cleaning the fields (9%) and applying salt (2%). The traditional methods (*cara kampong*) included using kitchen or padi husk ash, planting a branch of the mock willow tree (*Sapium indicum*) or a piece of bamboo painted red and scattering branches of *Lantana* in the field (Ho, 1983). Most (86%) sought advice when they encountered the problem either from MADA (93%), relatives and friends (6%) or the Department of Agriculture (1%).

When asked what steps they would take if their fields were attacked by *penyakit merah* in the next season, 79% said that they would use insecticides (Figure 20.1). If their neighbors' fields were attacked, 55% would also use the same insecticides, or use fertilizer (4%), destroy diseased plants (2%) or do nothing (9%). The remaining 29% gave a range of responses including informing the neighbor whose field was attacked, informing MADA offices, checking fields to control early, drying his field, organizing community help and offering advice to the neighbor.

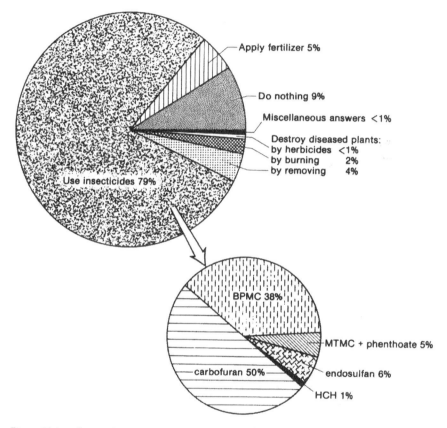

Figure 20.1   Farmers' responses to potential penyakit merah problems (% adopting various tactics)

When asked to speculate on the possible causes of the *penyakit merah* epidemic, 54% of the farmers did not know. Other responses included staggered planting (17%), being unable to dry fields (10%), high GLH populations (9%).

A range of control strategies for *penyakit merah* was suggested by farmers, including using insecticides in the nurseries (23%), detecting the problem and controlling it early (19%), clearing the fields (10%), drying the fields (5%), growing resistant varieties (5%), synchronous planting (4%), organized community help (2%), using less fertilizer (1%) and removing diseased plants (1%).

Effect of the Penyakit Merah Control Campaign

Of the 249 farmers interviewed, 18% had not heard of the campaign and 28% had not attended a single campaign briefing or *ceramah* organized by MADA; 51% had attended the ceramah at least three times during the campaign period and 28% had attended once or twice. Most (93%) felt that the information provided was useful.

The leaflets and posters prepared by MADA for the campaign reached about 80% of the farmers and most were aware of these from the MADA offices. Others had seen them in mosques, farmers' associations, shops and community halls.

During the campaign, farmers were urged to destroy diseased crops by burning after harvest, spraying herbicides (especially in water-logged areas) or hand weeding. Most (79%) agreed that such practices were effective and practical.

Farmers were also asked if they would be willing to participate in future MADA-organized control campaign briefings but only 49% said that they would, probably because they felt that no additional information was added at these campaign briefings.

When asked whether they would agree to MADA's strategy of not releasing water and allowing fields to dry for the month of February each year, to prevent *penyakit merah* epidemics, 90% said yes. Among those who did not agree, the reasons given were "farmers would lose income" (63%), "farmers would be forced to plant late" (13%) or that the strategy would not work (25%).

DISCUSSION

Penyakit merah in Bahasa Malaysia means 'red disease' referring to plants with yellow to brown discolorations. These symptoms may be caused by iron toxicity, organic soils, bacteria, insects (the white back planthopper), fertilizer deficiency, water stress or phytotoxicity caused by herbicides. While distinguishable by scientists, farmers are likely to classify them all as one problem, describing the symptoms as *penyakit bom* or bomb disease because of its sporadic occurence and rapid spread in rice fields.

Most farmers in the study could recognize *penyakit merah* symptoms and GLH as their cause, probably through the activities of the MADA campaign. However, they apparently did not perceive the virus as the causal agent. This lack of understanding of diseases is also found among Filipino rice farmers (Litsinger et al., 1982).

A large proportion of the farmers has adopted chemical strategies against *penyakit merah*, probably because they perceived the problem to be caused by insects. The campaign activities had recommended the use of carbofuran granules and BPMC sprays as prophylactic treatments for tungro control and perhaps influenced farmers' decision making. There was an increase in farmers' use of insecticides, compared to 1981, when only 62% used insecticides (Heong et al., 1985). The currently recommended chemicals are more effective than the endosulfan and HCH used in 1981.

The campaign also emphasised the use of cultural methods and resistant varieties but they were not widely adopted, perhaps because farmers were not confident that these methods alone would suffice. Given farmers' commitment to insecticide-based strategies, and the lack of awareness of the actual cause of the disease, future control campaigns and extension should place more emphasis on non-insecticide methods of control.

When their neighbors' fields were attacked by *penyakit merah*, more than half the farmers interviewed said that they would protect their crop using insecticides. This risk averse behavior is probably typical of small farmers (Norton and Conway, 1977). The farmers spend little time on their farms, and they are probably more risk averse to pests and diseases, adopting prophylactic control measures. Wong (1983) found that most farmers in the Muda spent less than 15 full working days per season on the farm and, of this, less than 1% was spent on crop protection. The other actions mentioned by the farmers, such as informing the neighbors and informing MADA, suggest a high community spirit among the farmers. Organizing community labour exchange schemes or *berderau* during periods of high labour demands is an example of this. Wong (1983) found that a large proportion (24%) of the farmers participated in such labour exchanges during the periods of transplanting and harvesting.

The *penyakit merah* campaign had reached most of the farmers in the Muda. The fact that so many had attended the *ceramah* more than once may be due to the insecticide handouts given at the end (Samsudin, 1984, pers. comm.). Leaflets and posters were also effective components of the control campaign principally via MADA offices and farmers' associations. This is probably because these are the distribution centres for the fertilizer subsidies and farmers are likely to visit them at least once a season.

During the campaign, farmers were urged to apply herbicides to ratoon crops, remove diseased plants by hand and dry plough immediately after burning the straws. Although these practices initially met with some resistance, they became quite widely accepted after a short period. The increase in dry-ploughed areas (Table 20.1) may also be due to the increase in farmers' adoption of direct seeding for crop establishment (Ho, 1983).

In 1984, the MADA implemented a water management policy of instituting a fallow period of one month, in February each year (MADA, 1984) when irrigation water is shut off. The policy was adopted to conserve irrigation water, improve crop scheduling, maintain the padi soil bearing capacity and improve pest and disease management. Since its implementation, penyakit merah occurences in the Muda have been reduced, so it is not surprising that 90% of the farmers interviewed were in favor of the fallow period. Penyakit merah has indirectly made it easier to introduce this stringent water management strategy.

ACKNOWLEDGMENTS

The authors are grateful to the General Manager of MADA, Dato' Syed Ahmed Almandali, for his support in the project, to Mr. C.P. Sau, Mr. Darus and Mr. Mahendra of MARDI who helped in the data processing, to the field technicians of MADA who assisted in the project and to the Consortium for International Crop Protection (CICP) and Dr. Ray Smith for support for one of us (KLH) to attend the PMPP Conference.

REFERENCES

Afiffuddin, Hj. Omar (1977a). Irrigation structures and local peasant organization. MADA Monograph No. 32. Muda Agricultural Development Authority, Alor Setar, Kedah, Malaysia.

Afiffuddin, Hj. Omar (1977b). Some organizational aspects of agricultural and non-agricultural growth linkages in the development of the Muda region. MADA Monograph No. 33. Muda Agricultural Development Authority, Alor Setar, Kedah, Malaysia.

Heong, K.L. (1984). Pest control practices of rice farmers in Tanjong Karang, Malaysia. *Insect Science and its Application*, 5: 221-226.

Heong, K.L., Ho, N.K. and Jegatheesan, S. (1985). The perception and management of pests among rice farmers in the Muda Irrigation Scheme, Malaysia. Malaysia Agricultural Research and Development Institute Report No. 105.

Ho, N.K. (1979). The framework of agricultural extension programs in the Muda Scheme — A quick glimpse. MADA Monograph No. 39. Muda Agricultural Development Authority, Alor Setar, Malaysia.

Ho, N.K. (1983). Traditional methods of pest control in the Muda area. Malaysian Plant Protection Society Newsletter, 6: 3-4.

Johnston, A. (1954). Preliminary notes on physiological disease of rice in Malaysia. FAO International Rice Commission Newsletter. No. 10.

Lim, G.S. (1969). The bionomics and control of *Nephotettix impicticeps Ishihara* and transmission studies on its associated viruses in West

Malaysia. Bulletin No. 121, Ministry of Agriculture and Cooperatives, West Malaysia.

Lim, G.S., Ting, W.P. and Heong, K.L. (1974). Epidemiological studies of tungro virus in Malaysia. *Malaysian Agricultural Research and Development Report* No. 21.

Ling, K.C. (1972). *Rice Virus Diseases.* International Rice Research Institute, Los Banos, Philippines.

Ling, K.C., Tiongco, E.R. and Flores Zenaida, M. (1982). Epidemiological studies of rice tungro. In *Plant Virus Epidemiology*, eds. J.M. Thresh and R.T. Plumb: Blackwell Scientific Publications, Oxford, England. pp 249-257.

Ling K.C., Tiongco, E.R. and Cabunagan, R.C. (1983). Insect vectors of rice virus and MLO-associated diseases. In *Proceedings of the 1st International Workshop on Leafhoppers and Planthoppers of Economic Importance.* Commonwealth Institute of Entomology, London, pp 415-437.

Litsinger, J.A., Price, E.C. and Herrera, R.T. (1980). Small farmer pest control practices for rainfed rice, corn and grain legumes in three Philippine Provinces. *Philippine Entomologist* 4: 65-86.

Litsinger, J.A., Canapi, B. and Alviola, A. (1982). Farmer perception and control of rice pests in Solana, Cagayan Valley, a pre-green revolution area of the Philippines. *Philippine Entomologist* 5: 373-383.

Matteson, P.C., Altieri, M.A. and Gagne, W.C. (1984). Modification of small farmer practices for better pest management. *Annual Review of Entomology* 29: 383-402.

MADA (1984). Technical considerations in the management of cropping schedules in the Muda Irrigation Scheme. Muda Agricultural Development Authority General Manager's Office, Alor Setar, Kedah, Malaysia. Mimeo.

Mumford, J.D. (1981). Pest control decision making: Sugar beet in England. *Journal of Agricultural Economics* 32: 31-41.

Mumford, J.D. and Norton, G.A. (1984). Economics of decision making in pest management. *Annual Reviews of Entomology* 29: 157-174.

Norton, G.A. (1982). A decision analysis approach to integrated pest control. *Crop Protection* 1: 147-164.

Norton, G.A. and Conway, G.R. (1977). The economic and social context of pest, disease and weed problems. In *Origins of Pests, Parasite, Disease and Weed Problems* eds. J.M. Cherett and G.R. Sagar. Blackwell Scientific Publications, Oxford, pp 205-226.

Ou, S.H., Rivera, C.T., Navaratnam, S.J. and Goh, K.G. (1965). Virus nature of penyakit merah disease of rice in Malaysia. *International Rice Commission Newsletter* 15: 31-33.

Prasadja, I. and Ruhendi (1980). Farmers existing technology and pest control practices for food crops at three locations in Yogkarta Province. Report from Agency for Agricultural Research and Development, Central Research Institute for Agriculture, Bognor, Indonesia.

SAS (1982). *Statistical Analysis Systems User Manual*. 1982 edition.

Tait, E.J. (1978a). Factors affecting the usage of insecticides and fungicides on fruit and vegetable crops in Great Britain: I. Crop-specific factors. *Journal of Environmental Management* 6: 127-142.

Tait, E.J. (1978b). Factors affecting the usage of insecticides and fungicides on fruit and vegetable crops in Great Britain: II. Farmer-specific factors. *Journal of Environmental Management* 6: 143-151.

Ting, W.P. and Paramsothy, S. (1970). Studies on penyakit merah disease of rice. I. Virus-vector interaction. *Malaysian Agricultural Journal* 47: 290-298.

Thompson, A. (1940). Notes on Plant Diseases in 1939. *Malaysian Agricultural Journal* 28: 402.

Wong, H.S. (1983). Muda II Evaluation Survey — An Impact evaluation study of the Muda II Irrigation Project. Muda Agricultural Development Authority, Alor Setar, Kedah, Malaysia. Mimeo.

21

Pesticide Practice Among Vegetable Growers In Mauritius

I. Fagoonee

INTRODUCTION

There is growing worldwide concern about environmental conservation and the preservation of wildlife, and at the same time there are persistent calls for increased food production. The latter demands more effective control of insects and other pests which threaten human health, livestock and crops, and as a result, increasing use of chemical pesticides. Balancing environmental protection and intensity of pesticide usage is a difficult and sensitive issue. Rational pesticide use is one way of decreasing dependency on pesticides, but its success requires among other things the collection of baseline data on pesticide strategies and on the perception of pests and pesticides among those involved in crop protection.

Studies have been initiated on pest and pesticide strategies in Mauritius. The status of economically important pests and their control, pesticide imports, pesticide legislation and management, and aspects of pesticide usage have already been reported (Fagoonee, 1984a; b). Following surveys to obtain a profile of national pest and pesticide management strategies a national Pesticide Management Advisory Committee was set up (Fagoonee, 1984a) to complement and strengthen the activities of the existing Pesticides Control Board. This report updates previous findings and is based on a national survey of pesticide usage and associated safety aspects, perceived pest problems and sources of advice.

PESTICIDE USAGE IN MAURITIUS

There has been a steadily rising trend in the import of pesticides to Mauritius over the past decade as indicated in Table 21.1. However, total food crop production (40,000-50,000 tons) and the acreage cultivated (about 10,000 acres) has not changed substantially. The synthetic pyrethroid insecticide, deltamethrin (Decis), was introduced in 1979 and is now the most widely-used insecticide on tomatoes, potatoes and crucifers, despite increasing signs of resistance.

A disturbingly large number of pesticides (over 40 insecticides, 20 fungicides and 30 herbicides), some of which have been banned or

TABLE 21.1

Pesticide imports to Mauritius 1970-1981 (tons formulated pesticide, three year moving average)

Year	Insecticides	Fungicides	Herbicides
1970-72	148.5	37.3	544.2
1971-73	158.6	49.4	464.3
1972-74	171.1	55.0	500.9
1973-75	180.1	61.1	544.0
1974-76	186.1	62.6	578.1
1975-77	262.6	61.5	562.4
1976-78	288.3	58.3	620.4
1977-79	353.5	72.2	658.8
1978-80	380.2	72.1	732.4
1979-81	374.1	85.3	709.3

Source: Annual Reports, Customs and Excise Department, Mauritius, 1969-81.

restricted in developed countries, is imported to the island. Parathion was introduced in 1949, trichlorphon and diazinon in 1952, azinphosmethyl in 1953 and dimethoate and endosulfan in 1956 (Galowalia, 1973). DDT has also been imported since 1948.

There has been an alarming increase in the number of cases of accidental and suicidal poisoning by pesticides, responsible for 80-90% of all poisoning cases in Mauritius. There were 90 deaths in 1979, compared to 24 in 1971. Endosulfan (11 deaths in 1977, 15 in 1979), dimethoate (8 in 1977, 10 in 1979) and paraquat (7 in 1977, 9 in 1979) are the pesticides most frequently incriminated (Fagoonee, 1984b). Recently the monitoring of blood cholinesterase levels has done much to improve the health of workers most exposed to pesticides, the percentage of serious poisoning cases decreasing from 35 in 1976 to 2 in 1981.

SURVEY METHOD

There were 2,837 farmers in Mauritius in 1983, divided into 25 agricultural zones and 207 localities for administrative purposes. Twenty localities, chosen at random, were sampled and approximately ten farmers from each (193 in total) were interviewed.

The questionnaire was designed in the local French-Creole dialect and the survey was carried out by extension officers of the Ministry of Agriculture in their respective areas. Data collected included age and level of education, size of holding, crops grown, purchase of pesticides, storage, advice, decision making, use of pesticide mixtures, health hazards, safety periods, spraying practices, disposal of empty containers and sprayer characteristics.

Where possible, the results are compared with the findings of two previous surveys (Fagoonee 1984a; b), a national survey of small scale vegetable growers in 1979 and a survey in 1980 covering 73% of all vegetable planters (60) in a coastal village.

RESULTS AND DISCUSSION

The growers in the survey reported here were, on average, younger and better educated than those in the earlier surveys. Nineteen percent had completed secondary education and 3% had university degrees, while the previous survey had none in these categories. The percentage with no schooling at all had also declined from 23% in 1979 to 19% in 1983. Tomatoes and potatoes occupied more than one-third of the total acreage devoted to food crops, followed by chillies, beans, cucurbits, groundnut, pineapple, cabbage, eggplant, onion and cauliflower, together representing over 56% of the total acreage (Table 21.2). Other crops grown, on a smaller scale, included ginger, carrot, garlic, pea and squash. Forty percent of farmers grew tomatoes and only 6% grew potatoes, the latter being grown on larger holdings.

TABLE 21.2

Distribution of major crops grown by farmers in the sample

| Crop | Area | | Growers | |
	Acres	% †	No.	%
Tomatoes	72	20.6	77	39.9
Potatoes	49	14.0	12	6.2
Chillies	33	9.4	51	26.4
Beans	22	6.3	36	18.6
Cucurbits	22	6.3	29	15.0
Groundnut	22	6.3	13	6.7
Pineapple	22	6.3	5	2.6
Cabbage	21	6.0	40	20.7
Eggplant	20	5.7	30	15.5
Onion	19	5.3	23	11.9
Cauliflower	11	3.1	22	11.3

† % of total crop acreage surveyed.

Pest Recognition

Major pests and diseases recognized by farmers were as follows:

i) 56% claimed to recognize caterpillars, including all Lepidopterous larvae, namely *Heliothis armigera* (fruit borer), *Phthorimaea operculella* (tuber moth), *Spodoptera litura* and *Spodoptera* spp, other cutworms and cruciferous pests *Crocidolomia binotalis* and *Plutella xylostella*;

ii) 41% recognized leafminers, mainly *Liriomyza trifolii*;

iii) 20% recognized red spider mite;

iv) 17% recognized fruitflies, mainly *Dacus cucurbitae* and *D. ciliatus* on cucurbits, *Paradalaspis cyanescens* on potatoes and tomatoes, *Ceratitis capitata* (Mediterranean fruitfly) and *Pterandus rosa* (Natal fruitfly) on chillies;

v) 13% recognized blight, due to *Phytophthora infestans* and other fungi.

Other pests mentioned by a few farmers were mites, aphids, mealy bugs, pod borers, leaf tiers (folders and rollers), slugs and snails, cutworms and thrips. Sixteen percent of farmers did not recognize any pests.

Most planters were not capable of putting a specific name to a pest. Among the caterpillars, a few farmers could distinguish between pod borers, cutworms and leaf tiers but otherwise they used the collective term 'caterpillar'. The commonest pests reported by most farmers were leafminers, fruit flies and caterpillars.

Pest Control and Pesticide Use

So far pesticides are the only solution to crop protection problems. The most popular pesticide is deltamethrin (Decis), used by two-thirds of the farmers, followed by methamidophos (Tamaron) and mancozeb (Dithane M-45). The use of cypermethrin (Cymbush) and chlorpyrifos (Dursban) is increasing due to a reported decline in the efficacy of deltamethrin, possibly due to resistance. Table 21.3 shows the pesticides used by farmers on the major crops grown, a wide range of pesticides being used on several crops.

Acquisition and Storage of Pesticides. Only 36 of those interviewed kept a record of pesticide purchases. For advice on pesticide purchases, 91% of those surveyed used the extension officer, 54% used salesmen, 42% used retailers and 23% used neighbors. Those who did not feel they needed advice (4%) claimed to have acquired their skill through experience and education. In the 1979 surveys, only 71% needed advice, 54% using their neighbors and 13% using extension officers (Fagoonee, 1984b). Storage of pesticides was more satisfactory than in the previous surveys. Most farmers used locked (66%) or specially assigned rooms or cupboards (38%) for storing pesticides. They also seemed more aware of suicidal and accidental pesticide hazards.

TABLE 21.3

Pesticide usage on major crops (% of farmers using each chemical)

Pesticides	Crops †										
	Tm	Pt	Ch	Be	Cc	Gn	Pn	Cb	Ep	On	Cf
deltamethrin	65	67	8	61	28	8		60	20	17	64
methamidophos	31	58	12	19	14	23		38		48	32
mancozeb	40	58	4	25	28	15		30	13	43	36
methomyl	1				21			5			5
monocrotophos	3			3	3	23		3			5
cypermethrin	5			6				8			9
omethoate	3	8	2				80		10		
amitraz	5		6						27		
quinomethionate	1		20						7		
trichlorphon	4				10			3			5
dimethoate			4	3				3	10		5
binapacryl	4		4					,	7		
phosphamidon			2				20				
parathion				3						22	
azocylotin			4						7		
sulfur			8	3							
metalaxyl	5	50									
diazinon		8									5
carbofuran		8	2								
others ‡	3		2	3	7			3			5
Total no. of pesticides	13	7	13	9	8	4	2	9	8	4	10

† Tm=Tomatoes; Pt=Potatoes; Ch=Chillies; Be=Beans; Cc=Cucurbits; Gn=Groundnuts; Pn=Pineapple; Cb=Cabbage; Ep=Eggplant; On=Onion; Cf=Cauliflower

‡ including benomyl, endosulfan, fenitrothion, fenthion, phenthoate.

Application Rates. Advice on application rates was obtained from various sources: extension agents (80%), experience (52%), labels (51%), salesmen (41%) and neighbors (18%). Schedule spraying was adopted by 65%, 37% sprayed when damage was seen, 35% sprayed even when damage was not seen, 30% when advised, and 7% only when they were able to procure pesticides. To measure pesticides, 92% (89% in 1979) used any sized spoon, 23% (2% in 1979) used a measuring cylinder, 6% used a can, 10% used a scale pan, and 27% (14% in 1979) used 'the dose' (i.e. the amount contained in a packet or a can for a given volume of water or for a given acreage). A wide range of dose-rates was applied on some crops, sometimes varying by a factor of five.

Use of Pesticide Mixtures. There was strong tendency towards the use of pesticide mixtures as follows: 1 insecticide + 1 fungicide, 85% of farmers; 2 insecticides, 4%; 2 insecticides and 1 fungicide, 35%. In addition 89% applied fertilizers or other additives such as sticker (48%), growth regulator (4%) or an attractant (0.5%), along with pesticides.

Recommended Safety Period. Only 9% of the interviewees observed the recommended safety period for a given pesticide: 9% would harvest one day after the last spray, 17% after two days, 69% after seven days; 29% after 14 days; and 14% after 21 days. Weekly harvests are regular features and biweekly harvests are becoming more common. There was a serious departure here from recommended norms.

Wearing of Protective Clothing. The situation regarding protective clothing was better in 1983 than in the 1979 survey: 86% wore rubber boots (38% in 1979); 64% wore rubber gloves (26% in 1979); 13% had a special overall; and 12% used face shields. Only 14% (62% in 1979) did not use any protective device.

Spraying Time. The preferred spraying times were: morning 62%; noon 50%; afternoon 23%; and anytime, 16%. Weather conditions that would prevent spraying were: wind and rain 70%; wind and cloud 31%; wind and sun 30%; calm and rain 48%; calm and cloud 36%, calm and sun, 9%.

Sprayers. The majority of the planters (98%) used knapsack sprayers, with 1% using motorised sprayers and 3% not using sprayers at all. In this survey most farmers washed out their sprayers after spraying — 97% compared to only 50% in the previous surveys. However, the number who washed out the sprayer in a river or canal had also increased — 30%, compared to 18% in 1979.

Disposal of Containers. The methods adopted for disposing of empty containers were: abandoned or thrown in bushes 62% (83% in 1979); buried or burned 41% (9% in 1979); re-sold to seller 27% (1% in 1979).

Health Problems. For the first time planters complained about burns (52%), headache (21%), vomiting (8%) and nausea (7%). Only 39% reported none of these symptoms.

CONCLUSION

There has been a net improvement in pesticide usage practices compared to the surveys done in 1979. Many more planters are using extension officers for advice and information. However, data on perceptions are still inadequate. Despite rigorous legislation on the import and sale of pesticides, officially banned or restricted pesticides are still available, possibly at lower prices, to those who prefer high toxicity, broad spectrum activity and low degradability, regardless of side effects.

Surveys such as those reported above should be improved in the light of experience gained, and should become a regular feature of crop protection activities to monitor growers' attitudes. Besides perception studies it is important, for comparative purposes, that pesticide usage data be standardized (Tait, 1977). Crop-specific studies, especially on the major crops, would indicate how to improve pest and pesticide management where pests are more prevalent and pesticides most intensively used. Resistance studies are also needed. A comprehensive national report on pest and pesticide strategies would then be available.

REFERENCES

Fagoonee, I. (1984a). Pests, pesticides, pesticide legislation and management in Mauritius. *Insect Science and its Application* 5 (3): 175-182.

Fagoonee, I. (1984b). Pertinent aspects of pesticide usage in Mauritius. *Insect Science and its Application* 5 (3): 203-212.

Galowalia, M.M.S. (1973). Pesticides uses and properties. In *Some Aspects of Insect Pest Control Chemicals. Revue Agricole et Sucritere de l'Ile Maurice* 52: 73-77.

Tait, E.J. (1977). A method for comparing pesticide usage data patterns between farmers. *Annals of Applied Biology* 86: 229-240.

22

Trends In Pesticide Usage In Uganda

E.M. Tukahirwa

INTRODUCTION

Pesticides will remain essential for the development of the agricultural and livestock industries in Uganda, as elsewhere, but it is generally accepted that they have a potential for harm. If improperly used, they can cause direct human poisoning, accumulate as residues in food and the environment or lead to the development of resistant strains of pests. These problems can arise from misuse of pesticides or over-reliance on them, particularly if the users are not conscious of these potential problems. In Uganda, there are already species of ticks which are resistant to toxaphene (Kitaka et al., 1970), and there are probably other pests resistant to other pesticides, which are yet undetected due to lack of proper monitoring. To prevent such problems, pesticide users must consciously avoid misuse and over-reliance on chemicals, and also be prepared to consider other alternative methods of pest management. This paper reports the initial observations from a study carried out in Uganda with the following objectives:

i)   to determine the extent of pesticide usage, i.e. the proportion of farmers that use chemicals and to what extent they rely on them.

ii) to assess the extent of the farmers' knowledge of pesticides, and how aware they are of the chemicals' potential for harm;

iii) to assess the potential consequence of misuse of pesticides on the environment;

iv) to examine the management options for improving the safe and efficient use of pesticides.

Study Area and Survey Sample

Kasese District in Western Uganda was chosen for this initial study because it is an important vegetable growing area for urban markets in Kampala and other towns in south-western Uganda, and agriculture there is comparatively efficient. This area is also one of the leaders in cotton production in the country. Preliminary observations have also been made at the Kibimba Rice Farm in the east and on a number of dairy farms in the south of the country.

Kasese District presides over the Mubuku Irrigation Scheme, a government project started over a decade ago to boost agricultural production.

The scheme covers 2,000 acres, divided amongst 150 farmers, the average holding being ten acres. The scheme uses the waters of the River Sebwe which flows from the nearby Rwenzori Mountains, through Kitogo Swamp, into Lake George and then Lake Edward. The two lakes have a delicate ecological balance (Beadle, 1974) but nevertheless support a comparatively large fishery (Morgan, 1972). Only horticultural crops and cotton are encouraged on the scheme, farming being predominantly commercial and very intensive, with some farmers having harvest-season cash turnovers of up to 1.5 M Uganda shillings (the country's average monthly income per head is less than Shs 10,000). All the farmers on the scheme are members of the local Nyakatonzi Cooperative Society, which is important for their procurement of farm inputs including pesticides.

Besides those on the irrigation scheme, Kasese has other so-called progressive farmers (about 8% of all the farmers in the district excluding those of Mubuku). All progressive farmers contacted were also members of the cooperative. Also enjoying the benefits of the cooperative, but in a rather special way, is the Mubuku Government Prison farm, a large farm of about 1,200 acres, engaged in commercial production of over 300 acres each of cotton and maize and with hundreds of livestock. All the farmers in these categories were using pesticides as part of their farm operations, albeit to different degrees.

The majority of farmers in the district are subsistence farmers who normally do not use pesticides except, in some cases, on cotton. All such farmers visited had no links with the cooperative except as an outlet for their cotton.

The Nyakatonzi Cooperative Society is an arm of the larger Uganda Central Cooperative Union, established by government to advise farmers on agricultural improvement techniques and to import farm inputs (such as tractors, hoes, pesticides, fertilizers). Inputs can be sold to cooperators more cheaply than the free market prices because of tax concessions from the government. In some cases the cooperative also exports the farmers' produce. For example, Nyakatonzi buys all the cotton in its neighborhood and this is processed in the society's own ginnery at Kasese and then exported. It also maintains a shop in the local town where farm inputs are sold to members and non-members alike, but at different prices.

The activities of the three categories of farmer are overseen by the government extension service, comprising professional agricultural cooperative officers and local chiefs.

METHODS

Visits were made to individual farms, and pest problems, their remedial measures and other issues relating to usage of pesticides were discussed with the farmers. To select the farms to be visited within the Mubuku irrigation scheme, a list of the farmers was obtained from the agricultural officer in charge of the scheme, and the name of every fifth farmer on the list noted. Interviews were held on these farms with either the owner or his foreman. No interview was held if neither of these could be found. A total of 18 farmers on the irrigation scheme were interviewed. Outside the scheme, a similar procedure was followed using the taxpayers' list from the local chief. However, when a farm was either too far away or inaccessible by car, the fourth or sixth farmer on the list was visited. A total of 24 farmers outside the irrigation scheme was visited. Most of these farms were small holdings, ranging from four to eight acres, except two which were about 20 acres each, and the prison farm which was much larger. All were engaged in mixed farming, i.e., growing crops and keeping livestock, usually cattle.

OBSERVATIONS

Extent of Pesticide Usage

When farmers were asked to give an estimate of the quantity of pesticides used, the information they gave was too sketchy to be useful, mainly because they kept no farm records. However, all farmers could remember how frequently they sprayed each crop, and the names of chemicals most commonly used. Table 22.1 illustrates the varied nature of the chemicals, frequency of use by farmers and target pests in the Mubuku irrigation scheme. When questioned about the threshold they used to decide on spraying a crop, 12 farmers in Mubuku (67%) said that they spray as soon as they see, or someone else reports, an infestation (irrespective of its degree), or when the officer in charge advises them to spray. The remaining six usually sprayed prophylactically, because they did not want to take chances waiting for an infestation.

None of these farmers in Mubuku could recall a season when spraying was not done, and some admitted that they were sometimes under pressure from other farmers to spray, in case their crops acted as a reservoir for pests. When such pressure reinforces the farmers' own anxieties to obtain high yields because of the capital investment they put into each crop, the consequence is a high propensity to spray by all farmers on the scheme. Given the amount of pesticide going into the soil and being washed by irrigation water back into the Sebwe River from all 150 farms year after year, there is a potential pollution problem. This propensity to spray is exacerbated by the ready availability of relatively cheap pesticides from the cooperative. All these farmers received regular visits from the government extension officers (an average of one visit in about two weeks) but the officers were apparently doing little to dissuade farmers from such a heavy dependence on chemicals.

Outside the Mubuku scheme agriculture is less intensive and even progressive farmers who have access to pesticides from the cooperative use them much less. Of the seven progressive farmers visited, only two (plus the prison farm) were using pesticides on crops other than cotton, namely maize and groundnuts. All the progressive farmers, however, used

TABLE 22.1

Frequency of spraying pesticides by Mubuka farmers against the major pests of their most valuable crops

Crop	Chemical Used	Frequency of Spraying	Target Pests
Cotton	permethrin	3-4 to maturity	aphids, whiteflies, mites, caterpillars, cotton stainers
	cypermethrin	3-4 to maturity	aphids, whiteflies, mites, caterpillars, cotton stainers
Groundnuts	phosphamidon	2-4 to maturity	aphids, mealy bugs
Tomatoes	phosphamidon	0-1 per week to maturity	whiteflies, bollworms, mites
	mancozeb	1-3 to maturity	tomato blight
Beans	phosphamidon	0-1 per week to flowering	aphids
Crucifers	phosphamidon	0-1 per week to maturity	aphids, whiteflies, caterpillars
Maize	DDT	2-4 to half grown	stem borers, army worms
Onion	fenitrothion	1-2 per week to maturity	thrips

pesticides on cotton and for the control of tick-borne diseases in cattle, either by dipping the cattle or spraying them. The frequency of spraying cotton by progressive farmers was comparable with that in the Mubuku scheme, but they did not mention any pressure from neighbors who, in this case, were engaged in subsistence agriculture. These farmers were visited by extension workers once a month on average.

Chemicals for tick control were obtained from two main sources, the cooperative and the government extension officers, but some farmers also purchased small quantities from chemical company salesmen. In 1984, the seven progressive farmers interviewed had between them used five different chemicals, in a variety of formulations and trade names, including amitraz (Taktic), chlorfenvinphos (Supona or tick grease), dioxathion (Delnav, Supamix or tick grease), toxaphene (Coopertox or Pfizertox) and formothion (Aflix). However, due to a lack of records, it was not possible to establish the amount of each chemical used. All seven farmers had spray pumps of their own and there were also cattle dips on the two farms with most cattle. Cattle were dipped or sprayed routinely about once a fortnight.

Information from the neighboring district of Bushenyi, as indicated in their 1983 cattle census, shows that 83% of cattle were regularly dipped or sprayed, using similar chemicals to Kasese. Even some of the poorer non-progressive farmers were joint owners of communal dips, indicating that tick control is an activity on which most farmers feel that they ought to use pesticides. As noted earlier, this has already led to the emergence of strains of ticks resistant to some pesticides, implying that since the early 1960s when tick control by chemicals became widespread in the country, other kinds of damage such as pollution may also have been caused. There is thus a need for carefully conducted scientific studies to assess the possibility and degree of environmental contamination with pesticides. The resulting information would be useful in evaluating the chemical approach to pest problems in the context of environmental conservation and health.

The 19 subsistence farmers visited used few pesticides to varying degrees. Seventeen farmers (89%) had a field of cotton where they could have applied chemicals, but only 11 (58%) reported using any (either permethrin (Ambush) or cypermethrin (Ripcord) purchased from the cooperative shop). None of these 11 owned a spray pump, depending on borrowed or hired equipment. When asked what motivated them to spray, eight farmers responded with reasons such as "we are told that yields will be higher" or "the chiefs insist that we spray" or "the big farmers do it and therefore it must be worth doing". Only three farmers reasoned that they spray because sprayed cotton is usually of better quality and fetches more cash (Shs 90/kg compared with Shs 40/kg for poorer quality cotton; prices of agricultural produce are always under review, but this margin is usually maintained.) Only one of these farmers had been visited on the farm by an extension officer, and this was once, at the beginning of the planting period, after a public meeting during the officer's "grow more cotton" campaign. These responses suggest that, for this category of farmer, the motivation to spray an insecticide is not the urge to maximize income. They are thus less likely to become habitual pesticide users, and therefore, are unlikely to have much impact on the environment as far as pesticides are concerned. The higher unsubsidized prices of chemicals for non-cooperators probably act as disincentives to the use of pesticides by these farmers.

The eight subsistence farmers who did not use any pesticides on their crops said they could not afford the investment, even on cotton. Their cotton was usually intercropped with maize and beans, and their yield expectations were based mainly on fatalism. They were not unaware of the benefits of pesticides, as some of them bought insecticides (DDT, malathion or lindane) to mix with beans and maize to control storage pests. It is the author's view that farmers in this category lack the motivation to improve their production techniques, and need to be encouraged to use pesticides, among other things, to help them boost agricultural output, and to improve their quality of life. These remain the poorest of the farmers.

Kibimba Rice Farm

The possibility of pesticide pollution from irrigation schemes is illustrated further by Kibimba Rice Farm. This uses the waters of the River Kibimba which flows into the River Mpologoma and on into Lake Kyoga. Here, routine spraying starts with fallow land which is treated with herbicides, currently glyphosate (Roundup) or dalapon. When the land is cultivated, a pre-emergence herbicide (pendimethalin or oxadiazon) is applied, followed by a post-emergence application of bentazon or 2,4-D. From then on, up to six applications of insecticide and/or fungicide may be made before the crop matures. The list of these chemicals in store at the time of the visit included benomyl, futhalide, mancozeb, IBP (Kitazin), phosphamidon, dimethoate, fenitrothion, diazinon, trichlorfon, DDT, chlordimeform, EPBP and carbofuran. Moreover, the soil on the farm is reported to be so infertile that a good harvest cannot be expected without an application of fertilizer, several brands of which were being used by the farmers. With such a rigorous chemical routine for more than a decade, some river pollution is likely to have occurred. Besides fish and other aquatic fauna, this area is the home of numerous birds and other species. Unfortunately, as is the case at Mubuku, no environmental impact assessment is being conducted or is contemplated.

DISCUSSION AND CONCLUSIONS

The investigation shows that profit-motivated farmers, unlike those in subsistence farming, regard pesticide application as an investment. For farmers in Mubuku and at Kibimba Rice Farm, chemicals were depended upon to maintain high yields and ensure investment returns. Unless persuaded otherwise, these farmers will continue to use pesticides very liberally because they associate their success with pesticide usage. This certainly is true to an extent, but their success is also attributable to better overall farm management.

The need to moderate farmers' dependence on chemicals is enhanced by the fact that the irrigation waters flow back into rivers which are sources of livelihood for human communities, support varied animal and plant life, and also flow into lakes with important fisheries. Lake Kyoga in turn feeds the River Nile, and Lakes George and Edward are located in the Queen Elizabeth Park, which supports thousands of species of birds and other animals and plants, some of which are unique to this area. It would be an irony if wildlife in a national park established primarily for its preservation were allowed to deteriorate due to pesticide pollution from the Mubuku scheme.

The progressive farmers' urge to maximize crop income was somewhat less than those on the irrigation schemes, and they were using less pesticides on crops. They were, however, using large amounts of chemicals on cattle, which are still revered as symbols of wealth, irrespective of whether they bring in cash. Every effort is made to preserve their health, hence the widespread use of acaricides. As noted earlier, this has already caused the emergence of resistant strains of ticks and the need to switch to new acaricides. At current rates of usage, this situation is likely to be repeated, leading to an ever greater dependence on chemicals, therefore to further environmental pollution and an increased risk of contamination of human food supplies.

The information obtained on cotton growing has implications for the environment. Good quality cotton cannot be obtained without some pesticide usage and it is government policy to encourage its production and improve export earnings. This is also the policy towards the other export crops such as coffee, tobacco, tea, rice and vegetables. As production of these crops increases, usage of pesticides should be expected also to increase, especially among progressive and irrigation scheme farmers. In fact, according to Ministry of Finance Annual Reports, pesticide consumption in Uganda in 1982 increased by as much as 1800% over its 1981 level (Ministry of Finance, unpublished). Although there are no obvious indications of environmental pollution as a result of this increase in pesticide use, the risk is clearly there. A valuable course of action would be the introduction of an environmental monitoring program for identifying any evidence of pollution, or other unwanted side effects of pesticide use.

Tukahirwa (1984) stated that quantities of pesticides used in Uganda were not yet a cause for concern if properly managed. This study has shown that there is practically no danger of pollution from chemicals used by subsistence farmers who constitute about 90% of the farming community. Their plight is indicated in Figure 22.1 (a model based on farm visits by extension workers, extent of government subsidies on farm inputs, and supervision of farmers generally) which depicts the perceived relative importance of the various categories of farmer. The model shows that greater attention is usually given to more prosperous farmers.

Whereas it is justified that such farmers should receive all the advice and assistance they require, it is questionable whether this should be at the expense of the less productive, poorer subsistence farmers. The latter should be encouraged to use pesticides, if this will increase their productivity, thereby giving them enough to eat, and some excess to sell. However, usage of pesticides, even by subsistence farmers, should not be looked on as a panacea for the low productivity, poverty and hunger in Africa today. Long-term remedies to these problems lie in farming systems which are ready to employ alternative pest management options, and in

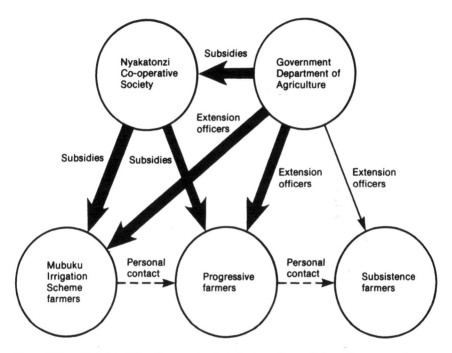

Figure 22.1   Model depicting the perceived relative importance of the various categories of farmers from data based on: farm visits by extension workers; extent of government subsidies on farm inputs; and supervision of farmers generally

Key: ➤ Government subsidies and extension officers' level of contact with different categories of farmer (not to scale); – – ➤ Low level informal contact between individual farmers government policies which are not discriminatory, but are supportive to the most vulnerable sections of society, especially in regard to subsidies and farm advice. There is also need for research on alternative pest management options to be incorporated into the training and planning for future agricultural activity. Present policies which tend to ignore research as an important input in agricultural development ought to be reviewed and a research component incorporated.

This paper is based on research carried out mainly in one region of Western Uganda. There is a need for more information to be collected from other areas of the country, particularly if such information were to be used in formulating national policies.

ACKNOWLEDGMENTS

I wish to thank farmers and officers of Government and Cooperative, for their cooperation and advice. Thanks are also due to Dr. J. Tait and Professor. R. Ford for their help and encouragement. This work was supported with a grant from the US General Services Foundation to the Committee for International Development and Social Change of Clark University, to which I am very grateful.

REFERENCES

Beadle, L.C. (1974). *The Inland Waters of Tropical Africa*. Longmans.

Kitaka, F., Oteng, A.K. and Kamya, E.P. (1970). Toxaphene-resistant ticks occurring on cattle in Uganda: *Boophilus decoloratus, Rhipicephalus avertsi* and *Rhipicephalus appendiculatus*. *Bulletin of Epizootic Diseases of Africa* 18: 137-142.

Ministry of Finance. (Unpublished). *Annual Reports Uganda Customs and Exise Department*, 1980 to 1982.

Morgan, W.T.W. (1972). *East Africa: Its Peoples and Resources*. Oxford University Press, Oxford.

Tukahirwa, E.M. (1984). Pesticides and their utilization: a profile for Uganda. *Insect Science and Its Application* 5(3): 183-186.

23

Farmers' Practices And Recommended Economic Threshold Levels In Irrigated Rice In The Philippines

H. Waibel

INTRODUCTION

In this paper farmers' crop protection decision making is compared to officially recommended practises. The data are based on research carried out under the Philippine-German Plant Protection Programme (PG-PPP) during 1979 to 1981 (Waibel, 1986).

The study areas were three pilot areas of the Regional FAO-IPC project in irrigated rice, in which pest monitoring was supported during the initial phase by the PG-PPP. The areas were located in three Philippine provinces, Nueva Ecija and Camarines Sur on the main island of Luzon, and Iloilo in the Western Visayas islands. Survey data was collected during the growing season and after harvest. Additional data are taken from trials undertaken by the Bureau of Plant Protection (BPP) and the International Rice Research Institute (IRRI).

FARMERS' PEST PERCEPTIONS

When investigating farmers' crop protection practices it is important to understand how they arrive at spray decisions. It is often assumed that they use calendar spraying, resulting in a high number of applications. In irrigated rice in the Philippines it was found that farmers, in one way or another, do monitor pests, before they make any spray decision.

As shown in Table 23.1, stemborers, mainly *Chilo suppressalis*, were the most frequently noted pests, followed by defoliators (*Spodoptera mauritia* and *Mythimna separata*), and the brown planthopper (*Nilaparvata lugens*), while the leaf folder (*Cnaphalocrosis medinalis*) was observed by only a few respondents. There was considerable divergence in the way farmers made observations and also deviations from the recommended sampling units. For example, in the case of stemborers 29% of farmers claimed to count tillers while others just looked at either the individual hill, an area not precisely defined or the appearance of the field. More farmers (75%) were prepared just to look at the crop, rather than count adult insects, leaves or tillers. This practice is reasonable from the point of view of labor, but is a questionable basis for spray decisions.

As shown in Table 23.2 the majority of farmers considered pest levels less than five as dangerous. This pattern was consistent across all major pests. Only 29% of responses indicated that the farmer would wait until the level reached ten or more. Table 23.3 shows that most farmers believed stemborers to be their major pest. However, stemborers were not abundant in fields monitored by experts, although they had been the major pest in the Philippines in the late sixties and early seventies. None of the respondents named leaf folders as their major pest, although this was frequently observed in monitoring plots.

The average number of applications of pesticide by the farmers in the survey was between one and three, depending on the area. A few farmers sprayed four times or more. We asked farmers cooperating in the pilot projects to quantify the yield losses they would expect in the absence of insecticides (Table 23.4). Only a minority estimated such losses to be below 10%. In two areas (Nueva Ecija and Iloilo) the majority of the respondents assessed yield losses to be within the range of 25 to 50%, and in one area (Camarines Sur) most assessments were placed at 50% or more. Combining the two upper intervals, the majority of farmers in all three areas assumed that they would loose more than one quarter of the potential yields if they did not use insecticides. Such high expectations for crop losses due to pests were not justified at the time when this study was carried out. Pest populations were generally well below the established economic threshold levels.

Farmers yield loss estimates were compared to data obtained from model calculations, based on pest data from untreated plots in combination with loss equivalents or loss functions. Figures were calculated for non-resistant varieties and varieties resistant to the tungro disease and the brown planthopper. While farmers' loss estimates ranged from 34% to 47%, depending on the location, those obtained from model calculations for resistant varieties ranged from 9% (Iloilo) to 13% (Nueva Ecija) (Table 23.5). This indicates that farmers over-estimate yield losses, given the present situation where resistant varieties are available. It also appears, that they base their estimates on negative events which happened in the past, probably during an outbreak situation.

Thus, farmers do observe pests, although they sometimes do it differently from the crop protection specialist. They interpret their observations in terms of 'dangerous' and 'not dangerous', but not in the recommended manner, and they expect high losses if they do not use insecticides. Given that farmers' behavior does not correspond with what is recommended, it is of interest to compare the economic performance of farmers' pest control decisions with those based on economic threshold levels for single pests.

TABLE 23.1
Criteria for farmers' pest observations (No. of farmers in each category, percentage in brackets)

Pest	Criteria of a pest observation counting			Criteria of a pest observation looking			Sampling criteria †
	Adult insect	leaf	tiller	hill	area	general appearance	
Brown Plant-hopper	6 (21)	—	—	10 (36)	—	12 (43)	No. per hill
Green Leaf-hopper	—	—	2 (50)	—	—	2 (50)	No. per sweep
Stemborers	—	—	15 (29)	19 (37)	10 (19)	8 (15)	Percent deadhearts
Defoliators	4 (13)	2 (7)	1 (3)	7 (23)	9 (29)	8 (26)	Percent damaged leaves
Leaf folders	—	—	—	—	1 (33)	2 (68)	Percent damaged leaves
Whorl maggot	—	—	—	—	2 (100)	—	Percent damaged leaves

† Sampling criteria of recommended economic threshold level

ECONOMIC ANALYSIS

Trials were carried out during three seasons on 58 farms, consisting of three treatments: farmers' practices; pesticide applications based on economic thresholds; and untreated plots. On average, pesticide treatments based on economic threshold levels (ETL) gave an *additional* net return of 145 Pesos per ha, compared to farmers' practices (Zeddies and Waibel, 1982). Compared to 'no spray' strategy, the margin was only 43 Pesos per ha on average, the benefit of the ETL strategy being mainly due to the reduction in the cost of control.

Where the ETL was reached, the ETL strategy performed only slightly better than farmer's practices, suggesting a fairly low control efficiency. However, looking at the individual cases, the ETL was reached in only seven out of 58 cases and in five out of the seven cases, ETL performed better than farmers' practices. Of the 51 cases, where the ETL was

TABLE 23.2

Levels of pest populations considered dangerous by farmers† (no. of respondents)

Pest	<5	5-10	10-20	>20
Brown planthopper	13	6	7	2
Green leafhopper	2	—	—	2
Stemborer	30	12	9	1
Defoliators	18	8	4	1
Leaf folders	1	1	1	0
Whorl maggot	2	0	0	0

† Criteria for observation not specified

TABLE 23.3

Farmers' assessment of their major pests in the wet season (% responding)

Pest	Nueva Ecija	Camarines Sur	Iloilo
Stemborer	55	52	42
Brown planthopper	—	4	43
Green leafhopper †	10	20	—
Defoliators	10	—	19
Whorl maggot	—	8	2
Rat	24	16	4
Others	2	—	—

† vector of tungro virus

not reached, there were 21 cases where the marginal revenue of farmers' practices was above marginal cost, i.e. the cost of control.

This means that, on average, farmers spray too much, but even so in about 40% of the cases they perform better than strategies based on recommended economic threshold levels.

The recommended ETLs are still not well defined. Partly this is because levels are based on single pests rather than on the most frequent pest combinations, and partly because the control efficiency of insecticide applications based on the ETL is rather low. Trials carried out by IRRI (IRRI, 1981) and trials performed in cooperation with BPP's regional crop protection centers, included treatments giving complete insect protection with about nine applications, ETL-based treatments and a control (Table

TABLE 23.4

Percentage expected yield losses due to pests in the absence of pesticides as assessed by farmers, wet season (% in each category)

Expected Loss (%)	Nueva Ecija	Camarines Sur	Iloilo
≤10	15	10	13
>10-25	23	10	30
>25-50	50	38	32
>50	12	42	25

TABLE 23.5

Yield loss due to pests as assessed by farmers in comparison to model calculations

	Nueva Ecija	Camarines Sur	Iloilo
Farmers' Estimates	34	47	40
Model calculation (resistant varieties)	13	13	9
Model calculations (non-resistant varieties)	40	38	15

23.6). Comparing potential loss with the loss after ETL-based treatments gives the control efficiency of the ETL strategy. On average, over five trials where only one pest reached ETL, there was a potential loss of 10.3%. The loss occurring after ETL-based treatments still amounted to 7.6%, resulting in a control efficiency of approximately 26%. This is not a convincing result for a technology which is recommended for small rice farmers who must strive for a minimum yield to maintain their level of living and whose behavior will therefore be risk averse. On the other hand if one expects benefits from economic thresholds in the reduction or avoidance of insecticide application, one faces serious limitations, in rice under the conditions observed.

Insecticides amount to between 3.8 and 6.2 percent of the total variable cost of rice production which would be about 100 to 150 Pesos, or about 5—8 US$ per ha. It may be difficult to convince a farmer to save this

TABLE 23.6

Control efficiency of recommended economic threshold levels for one pest based on results of field trials

Trial No.	Potential † Loss	Loss after ETL-based treatment	Control ‡ efficiency (%)
1	5.7	3.5	40.1
2	2.9	3.1	(0)
3	1.7	8.8	(0)
4	29.0	11.9	59.1
5	12.6	10.9	13.3
Mean	10.3	7.6	26.2

Source: Calculations are based on results of trials published in the IRRI insecticide evaluation report 1980 and own trials.

$$\dagger \quad \frac{\textit{Yield under maximum protection} \ - \ \textit{Yield under no protection}}{\textit{Yield under maximum protection}} \times 100$$

$$\ddagger \quad \frac{\textit{Potential loss} \ - \ \textit{loss after ETL-based treatment}}{\textit{Potential loss}} \times 100$$

amount of expenditure on insecticides and instead to follow rather laborious sampling techniques. This is only likely if there is a high yielding investment alternative for the money he can save by not spraying insecticides. IPM, therefore, should emphasize not only insect problems, but take into account the entire production process. Thus IPM would contribute to an optimal allocation of scarce farm resources like cash and labor rather than only aiming at reducing insecticide inputs.

CONCLUSIONS

The results reported here indicate that farmers observe pests rather than using calendar spraying, and that the pest levels at which they start spraying are generally lower than recommended ETLs. They also expect higher losses than actually occur, but do not use much insecticide. This gives them the impression that their pest control measures are very successful because the apparent benefit-cost ratio is quite favourable. However, the real benefit-cost ratio is not known to them. Economic analysis indicates that the real benefit-cost ratio for their crop protection practises looks quite favourable although the use of ETLs could help them to save small amounts of cash. To convince farmers' to follow recommended ETLs would require:

i) improvement of existing threshold levels by conducting trials in farmers' fields;

ii) improving farmers' knowledge of the yield loss to be expected from those pests abundant in the field. (Time should not be wasted teaching farmers about pests from other regions of the country).

Although there are no well established loss equivalents, farmers' over-estimates could be corrected by providing the necessary information. They could also be encouraged to leave an unsprayed part of the field to train them to carry out their own loss assessment.

ETLs are an essential part of integrated pest management and under the present situation for small rice farmers, the cash spent on insecticides competes with other input factors within the farm enterprise of rice production. IPM must focus on the optimal allocation of cash and also labor resources, rather than simply reducing the cost of control.

REFERENCES

IRRI (1981). *Insecticide Evaluation Report.* IRRI, Los Banos, Philippines

Waibel, H. (1986). The Economics of Integrated Pest Control in Irrigated Rice. In *Crop Protection Monographs,* ed. J. Kranz: Springer-Verlag.

Zeddies, J. and Waibel, H. (1982). Organization und Evaluierung eines Pflanzenchutzberatungs-projektes in einem Entwicklungsland. In 23. *Jahrestagung der GEWISOLA in Giessen.*

24

Perception And Management Of Crop Pests Among Subsistence Farmers In South Nyanza, Kenya

W. Thomas Conelly

INTRODUCTION

An increasing number of studies evaluating the potential of integrated pest management (IPM) have focused on socio-economic factors that may influence the successful introduction of improved methods of pest control among small-scale farmers in the tropics. Several researchers have studied farmers' perceptions of agricultural pests and yield losses, as well as traditional pest control practices. Research has also begun on socioeconomic constraints inhibiting the adoption of pest control practices recommended by national and international agricultural research centers (e.g. Adesiyun and Ajayi, 1980; Altieri, 1985; Atteh, 1984; Goldman and Omolo, 1983; Goodell, 1984; Heong, 1984; Litsinger et al., 1980; Matteson, 1984; Zaidi, 1984). Such research is vital for the development of IPM techniques that are technically feasible and appropriate to the circumstances of small-scale farmers.

More attention has been devoted to cash crop economies than to subsistence farming, where pesticide use is still restricted and alternative means of IPM may be more appropriate (but see Altieri, 1985; Matteson, 1984). In one such district, South Nyanza, Western Kenya, many farmers subsist by rainfed agriculture dominated by maize and sorghum production, using virtually no modern inputs. This paper presents the results of a study of existing pest management practices on maize and sorghum in South Nyanza focusing on insects, weeds, animals, and birds. Farmers' pest perceptions and management strategies, and the implications of this information for the development of a successful IPM package, are studied. Emphasis is placed on the broader context of the farming system and the feasibility of introducing two cultural control practices, the post-harvest destruction of the crop residues and early planting, that are potential components of an IPM package.

THE STUDY AREA AND METHODOLOGY

South Nyanza district, which extends from the shore of Lake Victoria, 1,128 m above sea level, to the foothills of the Kisii highlands, at over 1,500 m, is environmentally diverse. The rain fall distribution is bimodal, with a 'long' rainy season from March through May and a 'short' rainy season from October to December. In areas of low agricultural potential near the lake-shore, annual rainfall averages less than 900 mm and a single crop is produced in a year, most farming being subsistence-oriented. Farther inland at higher elevations, rainfall ranges from 1,200-1,800 mm per year, farming potential varies from moderate to very high and it is possible to harvest two crops each year with a surplus for the market

Population densities range from 60-80 per sq. km. near the western lakeshore to between 225-275 per sq. km. on Rusinga Island and in some of the high rainfall areas (Central Bureau of Statistics, 1981). The eastern area is well developed by Kenyan standards and is served by high quality tarmac roads providing access to major markets and government services. The western area is isolated and farmers' access to services and markets is restricted.

Information on farmers' perception and management of pests was collected through informal interviews and observations during the 1984 cropping season and by a formal post-harvest survey of a random sample of 48 farmers selected from three agroecological zones in the district (Figure 24.1): a) the low rainfall islands of Rusinga and Mfangano; b) Gera and Nyambunano in the low potential, near-lakeshore areas and c) in the medium to high potential inland areas Wiga and Koderobara. The majority of the data discussed in this paper is from the lower potential, single crop areas of subsistence farming.

The discussions with farmers took place in the local language, Dholuo, with the assistance of a translator. To assure the input of women who are responsible for much of the agricultural labour, the interviews were conducted with both the male and female heads of household (whenever possible) and eight of the interviews were with single women whose husbands had either died or were working outside the district. A few households were quite prosperous and the farmers 'progressive', but the majority were poor farmers, struggling to obtain the basic necessities of life. The estimated average farm size was 9.5 acres, ranging from 5.0 to over 15.0 acres.

FARMERS' PERCEPTION OF PESTS

To identify the pests that farmers considered to be threats to agricultural production, particularly to grain crops, we compared the perceptions of farmers with existing data on the distribution of insect pests in the district. Information on farmers' perceptions of pest hazards is critical in understanding the motives behind existing pest control practices and for identifying improved pest management techniques, appropriate to local

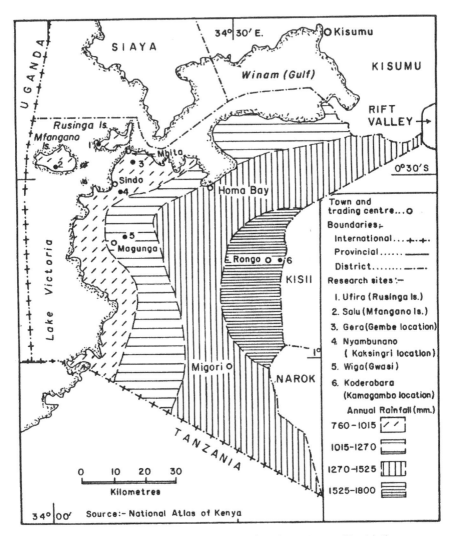

Figure 24.1   Research sites and mean annual rainfall – South Nyanza District, Kenya

circumstances.

The major insect pests of maize and sorghum have been identified as stem-borers, especially the spotted stem-borer *Chilo partellus* and the maize stalk-borer, *Busseola fusca*, and the sorghum shootfly, *Atherigona soccata* (ICIPE, 1984; Seshu Reddy, 1982). Initial damage from stem-borers occurs when the young larvae begin to feed on the leaves. Extensive tunnelling of the stem may occur and in severe cases plant growth is retarded and flowering and grain production are reduced (Teetes et al., 1983). Shoot-fly larvae enter the sorghum plant through the whorl and often destroy the growing point, resulting in a deadheart. Usually, the damage occurs from one week to one month after emergence. The sorghum plant may respond to shoot-fly attack by producing side tillers that are also susceptible.

Farmers were first questioned about the types of insect found on maize and sorghum. Stem-borers were the most common insect pest followed by white grubs, army-worm and termites, but the sorghum shoot-fly was not mentioned (Table 24.1).

TABLE 24.1

Percentage of farmers listing insect pests

Insect†	Islands n=16	Locality Lakeshore-Mainland n=16	Inland n=16	Total n=48
Stemborer	100	100	94	97
White grubs	44	0	6	25
Armyworm	0	13	13	8
Termite	6	19	0	8

† Identifications are tentative, based on local names, descriptions of insects, and discussion of crop damage given by farmers. A number of additional insects were described by farmers that cannot be identified with any confidence.

To measure farmers' ability to recognize common insect pests they were shown larval specimens of several stem-borers and the sorghum shoot-fly and asked to name the insect, the plant that it attacks, and describe the damage symptoms. Their ability to identify and describe the damage caused by *C. partellus* and *B. fusca* was impressive, with over 75% giving accurate descriptions. Many could give detailed information on the entry holes through which the larvae attack the plant, the tunnelling of the stalk which occurs, and the effect on yield. In contrast, though many farmers recognized the typical deadheart symptom of sorghum shoot-fly attack, none was able to correctly identify shootfly larvae.

Farmers' view of stem-borers as the major grain pest in the district is in agreement with field surveys. Failure to recognize the sorghum shoot-fly suggests that either the shoot-fly is not a serious pest in the area (contrary to survey data) or that farmers' knowledge of pests is faulty, perhaps due to the small size of the larvae, the fact that damage occurs very early in the development of the plant, or that the typical deadheart symptom can also be caused by stem-borers. This discrepancy requires further research.

Farmers were also questioned about other pests that affect their yields. An unexpectedly large number of wild animal and bird pests were identified as being common (Tables 24.2 and 24.3) and farmers could often provide extensive descriptions of their behavior and feeding habits.

TABLE 24.2

Percentage of farmers listing common animal pests

| | Locality | | | | | | |
Animal species	Rusinga n=10	Mfangano n=6	Gera n=8	Sindo n=8	Wiga n=8	Koderobara n=8	Total n=48
Hippopotamus	100	100	0	0	0	0	33
Wild pig	0	0	100	25	100	0	38
Monkey	0	100	100	25	0	13	35
Baboon	0	0	63	25	0	0	15
Porcupine	0	0	38	13	0	88	23
Mwanda†	40	0	25	38	38	75	38
Abur†	0	0	50	25	75	0	27

† Mwanda and abur are species of duiker and antelope that have not been positively identified.

Many farmers also said that weed pests were a serious threat to yields, particularly witchweed (*Striga hermonthica*), a parasitic weed that attacks the roots of maize, sorghum, millet, and occasionally sugar cane, especially on poor soils exhausted by continuous cropping (Ivens, 1968) (Table 24.4).

SEVERITY OF CROP DAMAGE DUE TO PESTS

Farmers' perceived severity of yield loss is critical in their decisions on the effort or expense to be put into controlling pests. Ideally, we would require a reasonably accurate estimate of the percentage of the crop lost due to each pest. Several researchers, using local methods of measurement that were later translated into percentages, have developed such estimates (Atteh, 1984; Zaidi, 1984). Unfortunately, in South Nyanza, insects are only one component of a complex set of constraints on production, and

TABLE 24.3
Percentage of farmers listing common bird pests

English name	Species	Locality Lakeshore n=32	Inland n=16	Total n=48
Weavers	*Ploceus spp.*	78	100	85
Dove	*Streptopelia spp.*	41	44	42
Seedeaters	*Serinus spp.*	25	50	33
Guinea Fowl	*Guttera edouardi*	19	6	15
Crowned Crane	*Balearica regulorum*	0	25	8
Crow	*Corvus sp.*	0	25	8

TABLE 24.4
Percentage of farmers listing common weed pests

English name	Species	Locality Lakeshore n=19	Inland n=16	Total n=45
Witchweed	*Striga hermonthica*	79	81†	80
Guinea fowl grass	*Rottboellia exaltata*	28	38	31
Couch grass	*Digitaria scalarum*	14	63	31
Blackjack	*Bidens spp.*	3	13	7

† Though *Striga* is present in the higher elevation areas, farmers report that couch grass is a more serious problem.

obtaining reasonably accurate estimates of yield loss due to any particular pest would be impossible. In some areas it would not be unusual to find a maize crop suffering simultaneously from several of the following problems: stem-borer damage, maize streak virus, *Striga* weed, soil erosion and drought. Assigning a percentage yield loss to each constraint would be a challenge even to an experienced agricultural scientist. Yet, if asked, most farmers will oblige the interviewer by providing a figure, no matter how inaccurate. If the pest situation in South Nyanza is typical, such estimates of yield loss may be more misleading than helpful, and figures should be viewed with skepticism, especially if they are used as a guide to policy recommendations.

As an alternative, in South Nyanza, we chose to ask farmers to rank various constraints to production, identified during the preliminary survey, in the order of seriousness of their impact on yields in most years. Overall, farmers' greatest concern was the risk of drought, followed by wild animals, weeds, and birds. Insects were ranked as the fourth most serious pest, followed by rodents. On the mainland, in both low and high rainfall environ-ments, insects were not considered a major constraint to grain production in most years. On Rusinga and Mfangano islands, however, insects were seen seen as a more serious threat to farming, out-ranked only by rainfall and animal pests (Table 24.5). This assessment corresponds with survey data showing relatively low insect pest infestation rates in fields located on the mainland away from the lake (e.g. Lambwe Valley and near Kisii border) but very high insect pest populations along the lakeshore, especially on Rusinga Island.

TABLE 24.5
Farmers' ranking of constraints to production — mean score (range 0-5) of each constraint, by location

Constraint	Islands n=16	Locality Lakeshore-Mainland n=16	Inland n=16	Total n=48
Rainfall	4.3	4.1	2.8	3.7
Animals	3.4	2.7	2.7	2.9
Weeds	2.5	2.6	3.1	2.8
Birds	1.8	2.5	2.7	2.3
Insects	2.7	1.8	1.8	2.1
Rodents	0.6	1.3	1.8	1.3

The lesson from this is that research ought not to be too narrowly focused on a single pest. Insects must be seen as only one of a complex of biological and physical hazards that limit yields, and farmers may be unwilling to adopt new farm practices designed to reduce insect pests if they ignore or exacerbate other serious agricultural hazards.

PEST CONTROL PRACTICES

A wide range of conscious pest control practices was employed against weed, animal, and bird pests. Recognizing the link between field fertility and *Striga* weed, a number of farmers used crop rotations and fertilization with manure to control it. Against animal and bird pests, farmers used an array of techniques such as guarding fields, scaring devices, fencing or destroying nesting sites.

Most farmers indicated that there was little they could do to limit insect pests. Traditional herbal 'insecticides' were remembered by some farmers, but have virtually disappeared from use. Others reported employing mechanical or physical controls, killing stem-borer larvae with a stone or hoe or uprooting heavily infested plants, but there was little evidence from observation or discussion with farmers that this was done systematically.

No farmers used insecticide on grain crops in 1984, explaining that the chemicals were unavailable, that they lacked the knowledge of how to apply them, or that they were too expensive. Some farmers did apply insecticide on vegetable crops destined for the market and insecticide is also widely used on cotton (Goldman and Omolo, 1983).

A number of farm practices may *unintentionally* control insect pests. For example, the intercropping of grain crops with legumes, though discouraged by the Ministry of Agriculture, is widely practiced and there is evidence that this helps to limit stem-borers (Amoako-Atta, 1983). However, only one farmer claimed that he intercropped because it controlled insects; 41% used intercropping because of labour or land shortages; 8% mentioned that it prevented weeds; 27% said that it gave a good yield without specifying a reason; 11% stated that it was traditional.

Thus, few conscious measures are taken by South Nyanza farmers to control insect pests in the field. Some authors (e.g. Atteh, 1984) suggest that a wide range of traditional agronomic practices, such as intercropping, have been adopted because farmers recognize that they help control insect pests, but this does not seem to be the case in South Nyanza. Most years, farmers in the higher rainfall areas such as Koderobara, do not see insects as a serious threat to production, warranting the use of pesticides or other control measures. In the low rainfall lakeshore areas such as Rusinga Island farmers apparently do recognize insects as a serious pest, yet take little action to control them. In these areas, where maize and sorghum farming is largely for subsistence, farmers may be willing to tolerate quite high losses to insects, given the high cost in cash and/or labour of control measures (Altieri, 1985). If farmers regularly suffer major yield losses to drought, as in lakeshore areas, and have alternative sources of subsistence and income such as livestock and fishing, then even substantial yield losses to insects may not justify the use of costly pest control measures (Goodell, 1984).

The willingness of some farmers to tolerate a significant reduction in yield as a result of insect attack suggests that it may be difficult to convince subsistence farmers to adopt IPM practices. New pest management strategies must be technically feasible and also offer a *significant* reduction in yield loss that will be worth the farmers' investment in terms of cash and labour expenditure. The greater use of insecticides in South Nyanza on cash crops such as vegetables and cotton suggests that small-scale farmers are less willing to tolerate damage by insects and other pests once their production is oriented toward the market. The same may be true for grain crops like maize when grown for sale, rather than home consumption (Goldman, 1987).

RECOMMENDED CULTURAL PRACTICES FOR CONTROL OF INSECT PESTS OF MAIZE AND SORGHUM

Two common recommendations for the cultural control of stem-borers and shoot-fly are: a) burning, after harvest, of the crop residue in which diapausing stem-borers are able to survive the dry season; and b) planting early in the rainy season to minimize the period during which vulnerable plants are exposed to high insect populations (Gahakar and Jotwani, 1980; Lawani, 1982; Seshu Reddy, 1982; Young and Teetes, 1977). The feasibility of such recommendations in the local farming system was investigated here.

As in other parts of Africa (Adesiyun and Ajayi, 1980) farmers in South Nyanza will hesitate to destroy crop residues after harvest because of the many uses for maize and sorghum stubble. Seventy-five percent of the farmers reported leaving stalks in the field as fodder for livestock during the dry season. Many others collected the stalks for fuel (48%) or for the construction of granaries (56%). In high rainfall areas farmers spread maize and sorghum stalks in banana and coffee orchards as a mulch. The crop residue also helps to limit soil erosion.

Thus alternative ways of controlling the carryover of stem-borers in the crop residue will need to be identified, e.g. partial burning of the stalks (Adesiyun and Ajayi, 1980) or delaying destruction of the stalks until the end of the dry season, when the remaining residue will have little value as fodder or fuel. Neither approach would be acceptable to farmers who use the residue as a mulch in banana and coffee orchards, often the most profitable aspect of the farm operation. In areas where burning may be economically justified, research is needed on the timing and amount of labour required to destroy the crop residue and the compatibility with existing farm practices.

The recommendation of early planting would also be difficult to introduce. Most farmers agree that, under ideal conditions, early planting is beneficial, but in practice it is often difficult to achieve, especially in the lakeshore areas. Many farmers argue that early planting increases pest problems. First, early planted fields are vulnerable to bird attack if they ripen before the fields of neighbors. Second, contrary to most scientific findings, many farmers feel that early planting increases rather than decreases the risk of insect damage. Sixty percent of the farmers said that they felt insect damage in their fields was higher in crops planted at the onset of the rains. Another 20% said that the insect damage was about the same for both early and late planted crops, but only 14% felt that early planting helped to reduce damage. Many farmers attributed the increase in insect damage in early planted crops to the erratic rainfall pattern. It is widely held that insect damage is only serious during periods of drought.

When the rainfall is regular and heavy, insects are said to be washed off the plant and drowned. Early rains are often followed by two weeks or more of light and uncertain precipitation, and many farmers felt that the newly germinated maize and sorghum was very vulnerable to insect attack at this time.

The erratic start of the rains also makes early planting risky, given the danger that seeds will germinate, but then fail to develop because of drought. In the serious drought in 1984, almost 70% of the farmers reported that they had to replant maize and/or sorghum at a prohibitive cost in seed and labour because of a false start to the rains. As a result of this uncertainty some farmers staggered their planting dates.

Early planting is further constrained in some areas, such as eroded hillside land on Rusinga Island and the black cotton soils found on the lakeshore mainland, because the soils are difficult to hoe or plough until they have been thoroughly moistened by rains. In 1984, when the first appreciable rains did not fall until March, farmers delayed planting until after the recommended date.

In addition to these environmental factors, socioeconomic variables such as unequal access to draft animals and farm implements can delay planting. In 1984, 14 farmers reported that they prepared their fields using only a hoe, which often resulted in delays in land preparation and smaller field size. Of the remaining 34 farmers who used a plough for land preparation, only 14 owned their own implements and plough animals. For the remainder renting or borrowing a plough and/or plough team often caused delays in land preparation and forced some to postpone their sowing date until several weeks into the rainy season.

CONCLUSION

IPM research should focus more on the unique circumstances of subsistence farmers who are often by-passed in the development of improved agricultural technologies. Low cost alternatives to chemical pest control must be identified and developed for these farmers. However, because of the major changes in labour allocation and resource use required by new cultural control practices such as the destruction of the crop residue and early planting, this component of IPM will be difficult to introduce into subsistence farming systems. Other components of an IPM approach, for example the development of resistant varieties of maize and sorghum, biological control, and inter-cropping (which is already widely practiced in South Nyanza) appear to be more promising alternatives to pesticide use. However, research on these aspects of pest management, cannot ignore the importance of key socioeconomic and environmental constraints that may influence the willingness and ability of subsistence farmers to adopt new IPM technologies.

ACKNOWLEDGMENTS

This research was conducted with the support of a generous social science research fellowship from the Rockefeller Foundation while the author was at the International Center of Insect Physiology and Ecology (ICIPE), Kenya. Special thanks to Abe Goldman at Clark University and to K. Ogada, A. Dissemond, J.K.O. Ampofo, E.O. Omolo, A. Alghali and M. Botchey at ICIPE for their interest and valuable assistance.

REFERENCES

Adesiyun, A.A. and Ajayi, O. (1980). Control of the sorghum stem-borer, *Busseola fusca*, by partial burning of the stalks. *Tropical Pest Management* 26(2): 113-117.

Altieri, M. (1985). Developing pest management strategies for small farmers based on traditional knowledge. *Bulletin of the Institute for Development Anthropology* 3(1): 13-18.

Amoako-Atta, B. (1983). Intercropping: a case study of the stem/pod borer interactions with maize cowpea sorghum cropping patterns in Kenya. ICIPE, Nairobi, Kenya.

Atteh, O.D. (1984). Nigerian farmers' perception of pests and pesticides. *Insect Science and its Application* 5(3): 213-220.

Central Bureau of Statistics (1981). *Kenya Population Census, 1979*. Ministry of Economic Planning and Development, Nairobi.

Gahakar, R.T. and Jotwani, M.G. (1980). Present status of field pests of sorghum and millet in India. *Tropical Pest Management* 26: 138-151.

Goldman, A. (1987). Agricultural pests and the farming system: a study of pest hazards and pest management by small scale farmers in Kenya. (This volume.)

Goldman, A. and Omolo, E.O. (1983). Farming practices and pest control in South Nyanza District, Kenya. ICIPE, Nairobi, Kenya.

Goodell, G. (1984). Challenges to international pest management research: do we really want IPM to work? *Bulletin of the Entomological Society of America* 30(3): 18-26.

Heong, K.L. (1984). Pest control practices of rice farmers in Tanjong Karang, Malaysia. *Insect Science and its Application* 5(3): 221-226.

ICIPE (International Center of Insect Physiology and Ecology) (1984). *Annual Report, 1983*. Nairobi, Kenya.

Ivens, G.W. (1968). *East African Weeds and Their Control*. Oxford University Press, Nairobi.

Lawani, S.M. (1982). A review of the effects of various agronomic practices on cereal stem borer populations. *Tropical Pest Management* 28(3): 266-276.

Litsinger, J.A., Price, E.C. and Herrera, R.T. (1980). Small farmer pest control practices in rainfed rice, corn, and grain legumes in three Philippine provinces. *Philippine Entomologist* 4: 65-86.

Matteson, P.C. (1984). Modification of small farmer practices for better pest control. *Annual Review of Entomology* 29: 383-402.

Seshu Reddy, K.V. (1982). Pest management in sorghum II. In *Sorghum in the Eighties,* Proceedings of International Symposium on Sorghum, 2-7 Nov., 1981. ICRISAT.

ibid (1983). Studies on stem-borer complex in Kenya. *Insect Science and its Application* 4(1/2): 3-10.

Teetes, G.L., Seshu Reddy, K.V., Leuschner, K. and House, L.R. (1983). *Sorghum Insect Identification Handbook.* Information Bulletin No. 12, ICRISAT.

Young, W.R. and Teetes, G.L. (1977). Sorghum shootfly entomology. *Annual Review of Entomology* 22: 193-218.

Zaidi, I.H. (1984). Farmers perception and management of pest hazard: a pilot study of a Punjabi village in lower Indus region. *Insect Science and its Application* 5(3): 187-201.

25

The Communication And Adoption Of Crop Protection Technology In Rice-Growing Villages In The Philippines

V.PB. Samonte, A.S. Obordo and P. Kenmore

INTRODUCTION

When a program of planned social change focuses on rice farmers, it is important to gain an understanding of their agro-social world and their linkages with the wider society, the context in which the process operates. In contemporary times, farmers have witnessed an explosion of agricultural technology and it is necessary to consider what agricultural technology has reached them on their farms, how this technology reached them, and how they have responded to it. This paper focuses on the communications that link rice farmers with sources of change. It identifies the key communicators who initially create awareness about farm innovations, the communicators who give instructions on the application of an innovation and those who bestow legitimacy on the acceptance of an innovation. The rice farmers, in turn, have to decide whether to adopt, reject or hold an innovation at bay. It is therefore necessary to identify the decision makers on the farm as this determines the target and nature of communications on farm innovations. Agricultural planners, policy makers, field implementors, extension agents and farm communicators need a basic understanding of the communication structures and decision making patterns of rice farmers at the village level before attempting to direct the uptake of crop protection technology.

This paper aims to outline the communication structure in rice-growing villages, identifying the communicators who create awareness, instruct and legitimate the adoption of crop protection technology, and to indicate the decision makers on the adoption of crop protection technology on rice farms.

METHOD

Data were collected using a combination of techniques, including an interview schedule, non-participant observation and sociometry at several entry points, from July to November 1984, during the rice cropping season.

On the basis of census and local data and consultative meetings with local officials of the Ministry of Food and Agriculture, four research locales were selected in Nueva Ecija, one of the primary rice producing provinces in the Philippines. The major criterion was accessibility. Rice farmers (107) were selected at random (Table 25.1).

TABLE 25.1
Survey sample distributions

Town	Village	(%)	Sample size No. of respondents
San Antonio	Panabingan (Village 1) (less accessible)	20	20
Paludpod	Talavera (Village 2) (less accessible)	100	33
General Natividad	Poblacion (Village 3) (accessible)	20	18
Cabanatuan	Caalibangbangan (Village 4) (accessible)	19	36

RESULTS

Agricultural Profile

Respondents were either landowners or certificate of land title holders, the majority (66-97%) tilling rice farms from 1-3.5 ha. The remainder had larger farms, from 3.6 to 8.5 ha. The farms were generally irrigated using combinations of pump, deep well, stream and rain water. Respondents, particularly in Poblacion and Caalibangbangan, augmented their basic earnings from rice farming by vegetable and livestock raising, income from other forms of employment and vending.

Adoption of Crop Protection Technology

A range of insecticides had been used during the previous two crop-ping seasons, the most common being monocrotophos, BPMC plus chlorpy-rifos, isoprocarb, methomyl, carbofuran, diazinon and methyl parathion. Hand pulling was mentioned to control weeds, but another frequently used control measure for weeds was the use of chemicals such as butachlor, 2,4-D, MCPA and piperophos. Rat infestations were generally controlled using poison or baits with zinc phosphide, coumatetralyl and warfarin, but rats that invaded the fields were also killed manually or physically. Rice varieties resistant to pests and diseases were often selected, particularly IR 36, IR 42, IR 58 and C1000. Most innovative practices were first intro-duced to the rice farms in the mid 1970s.

First Information Sources

In the adoption of new farm technology, some information sources serve specifically to make the potential adopter initially aware of the prac-tice. This function is performed by what is referred to here as the 'first information source' (Table 25.2).

In Village 1 (Panabingan, San Antonio), which was less accessible, the most frequently cited source was a Ministry of Agrarian Reform (MAR) technician. This technician created awareness for all five types of pest con-trol measures, although he was prominent in giving first information on insect and plant disease control. His central position may be partly explained by the fact that he lived in this village at the time when many of these innovations were first being introduced. Another key communicator for all control measures, except the selection of rice varieties, was the Bureau of Extension (BE) technician, who had worked in this village before being replaced by the present one. Among commercial personnel, represent-ing companies selling crop protection chemicals, one dealer was mentioned by some farmers, specifically for insect control measures. Mass media were mentioned to a lesser extent, particularly a radio program that broadcast information on weed control. Farmers were mentioned to some extent, especially for information on rat control and other resistant varieties.

In Village 2 (Paludpod, Talavera), also less accessible, the major first information source was the past BE technician who gave information on insect, disease, weed and rat control. Another past BE technician was the next most active communicator in this village. For the rest, this informa-tion function was dispersed among several sources with lesser frequency and for a lesser set of crop protection practices. For instance, the present BE technician gave information on how to control insects and plant diseases and on selection of resistant varieties. Some commercial dealers were first information sources for specific crop protection techniques, in one case insect and disease control, in another only for insect control or exclusively for weed control. Seven co-farmers served as lesser information sources, confined to a single crop protection technique. Radio programs 1 and 2 also gave out initial information on weed control.

TABLE 25.2
First information sources for crop protection on rice (% frequency of mention by farmers)

Source	Village			
	1	2	3	4
Insect control				
government technician	56	51	45	34
commercial personnel	21	16	32	35
co-farmers	20	28	19	16
mass media	3	5	2	14
others	—	—	2	1
Plant disease control				
government technician	82	41	68	46
commercial personnel	5	18	23	28
co-farmers	9	39	6	21
mass media	—	2	3	5
others	4	—	—	—
Weed control				
government technician	50	38	37	29
commercial personnel	7	11	29	33
co-farmers	18	31	29	24
mass media	11	15	5	8
others	14	5	—	6
Rat control				
government technician	50	37	50	27
commercial personnel	—	5	5	38
co-farmers	29	47	32	14
mass media	—	—	4	7
others	21	11	9	14
Rice varieties used				
government technician	13	35	62	12
commercial personnel	3	12	20	11
co-farmers	74	47	10	66
mass media	—	—	8	—
others	10	6	—	11

In Village 3 (Poblacion, General Natividad), an accessible rice farming community, the farmers depended mainly on both the past and the present BE technicians for initial information on all five types of crop protection technology, the former figuring prominently for insect and weed control, the latter for rice variety selection and plant disease control. Two farmers served as first transmitters of information, on insect, weed and rat control in one case and rat control only in the other. Other sources, limited to a single crop protection technology were: the Rural Bank Technicians for plant disease control; a commercial representative and Radio Program 3 for weed control; and a research project technician for selection of rice varieties.

In Village 4 (Caalibangbangan, Cabanatuan), also accessible, the present BE technician had the widest sphere of influence, covering all five areas of crop protection. The Land Bank technician was second in prominence for all crop protection technology except variety selection. Less important were a Rural Bank technician for insect, weed and rat control, and a farmer respondent for insect, disease and weed control. A Rural Bank technician and a commercial representative acted as sources for plant disease and rat control. For specific crop protection techniques, Rural Bank Stockholder and Radio Program 5 gave information on insect control and Radio Program 4 on plant disease control.

Additional Information Sources

Additional information sources included any source that imparted specific instructions on the procedures to be followed in using or applying a particular crop protection technology, performing an educational or teaching function on how to implement a crop protection method on the farm.

In Village 1, a MAR technician was most prominent, followed by the past BE technician who advised on all types of technology except variety selection. A farmer respondent shared his knowledge on rat control measures with some of the others. For variety selection, 13 different farmers were each reported by a respondent, indicating a dispersed communication pattern with the exception of the MAR technician mentioned above. Some respondents reported gaining knowledge through self-experience or self-study.

In Village 2, a past BE technician was the most popular additional information source for all five types of crop protection technology, particularly insect control. Also many respondents acquired knowledge through self-experience. A second past BE technician also enjoyed popularity as a teaching source for insect, plant disease and weed control. A third past BE technician had covered insect and plant disease control, and the present BE technician insect control and rice variety selection. Other sources such as commercial personnel, co-farmers and a MAR technician generally qave advice on only a single subject. For weed control, although there were favored sources, ten farmers were cited individually, indicating a partly dispersed pattern for this teaching function.

For accessible Village 3, the present BE technician was the most frequent source of instruction on the use of all five crop protection techniques, followed closely by the past BE technician. Two farmers were helpful in providing guidance on insect and weed control and on averting damage by rats. A rice research technician was cited as adviser on the selection of rice varieties, and a number of farmers referred to their own experience in this area.

The farmers in Village 4 most often relied on their own experience in applying crop protection practices. However, three advisers, one from the commercial and two from the government sector, were repeatedly mentioned as those who helped initially to familiarize them with crop protection techniques. These were the Rural Bank technician who provided instruction on insect control and rice variety selection, the Land Bank technician who gave guidance especially on plant disease and weed control and the BE technician assigned to this village. A farmer also disseminated additional information on all five areas of crop protection, particularly variety selection. One commercial representative supplied instructions on insect, weed and rat control while another commercial representative and two farmers covered insect control and variety selection.

Legitimation Sources

This communication function was defined as any source that gave information which served to persuade, influence or convince the respondents to decide in favor of adopting crop protection practices. Such sources pass on information that stamps approval, or lends support to the respondent in clinching his decision to use crop protection methods on his rice fields.

The prime legitimator in Village 1 across all five areas of crop protection technology, was a MAR technician. In many cases the respondents based their decisions on their own conviction that they were doing the right thing for their farms. To a lesser degree, the past BE technician gave persuasive messages favoring the adoption of insect and rat control methods. A farmer gave legitimacy to decisions on rat control methods.

The major legitimator in Village 2, across all five areas of crop protection technology, particularly insect control, was the past BE technician. Numerous respondents acted as their own self-legitimators, particularly in the selection of rice varieties and control of plant diseases. Legitimation was also performed to a lesser degree by the present and a past BE technician, a commercial representative, farmers and a MAR technician.

In Village 3, the legitimator who exerted the most influence in getting the farmers to adopt crop protection technology in all five areas was the present BE technician. Similarly, a past BE technician played a significant role in convincing the farmers to adopt measures to control insects, plant diseases, weeds and rats and to select good rice seeds for planting. A farmer proved to be the third most popular legitimator in this village. A commercial representative was instrumental in the acceptance of weed control practices. Several farmers served as their own legitimators in the adoption of insect and weed control and in variety selection.

In Village 4, the prime legitimators were the farmers themselves, whose own experiences convinced them of the importance of accepting these measures to ensure greater productivity. This was particularly evident in the case of seed selection. Advice from a Land Bank technician and a Rural Bank technician proved effective in making decisions to adopt crop protection practices, particularly disease control. Also two farmers had convinced a number of respondents to practice crop protection techniques, especially the selection of resistant varieties. The present BE technician's advice served to firm up decisions to accept appropriate technology to control insects, diseases and weeds, and the Rural Bank technician acted as legitimator for the control of rats and plant diseases.

Decision Makers on Crop Protection Practices (Table 25.3)

In Village 1 the primary decision maker was the farmer himself for all the five areas of crop protection technology. One MAR technician had a limited role in decisions to adopt control measures against insects, plant diseases and weeds on some farmers' fields.

Village 2 had a similar decision making pattern to Village 1. Here, four past BE technicians and the present BE technician performed this role to some extent, especially for insect control on rice fields.

In Village 3 farmers were also the dominant decision makers. However, there was an increase in the extent of joint decision making, involving the spouse, for technological control of insects and selection of rice varieties. Also the past and current BE technicians were cited to some extent as responsible for making decisions on insect and plant disease control.

The decision making pattern in Village 4 was relatively complex. Here, the primacy of the farmer as decision maker was still upheld but other decision makers were cited to a considerable degree, particularly the Land Bank technician for decisions on insect, plant disease and weed control, and Rural Bank technicians and other commercial personnel for decisions on insect control. A farmer was also active in this decision making role for weed and rat control and the selection of resistant rice varieties.

TABLE 25.3
Decision makers on crop protection practices (% of frequency of mention by farmers)

Decision makers	Village			
	1	2	3	4
Insect control				
government technicians	19	24	24	14
commercial personnel	—	2	—	17
self	73	72	53	49
others†	8	2	23	20
Plant disease control				
government technicians	18	13	27	26
commercial personnel	—	6	—	5
self	77	81	64	54
others	5	—	9	15
Weed control				
government technicians	15	16	8	25
commercial personnel	—	4	3	8
self	81	76	78	48
others	4	4	11	19
Rat control				
government technicians	4	16	24	13
commercial personnel	—	5	—	13
self	87	66	62	48
others	9	13	14	26
Rice varieties used				
government technicians	—	15	7	7
commercial personnel	—	—	10	6
self	85	72	53	73
others	15	13	30	14

† 'others' includes: government technician and commercial personnel, jointly; another farmer; farmer plus members of his family (plus government technicians), jointly; farmer plus co-farmers, jointly.

CONCLUSIONS

Generally speaking the farmers in Village 1 identified government technicians as their first sources of information for all aspects of crop protection technology except the selection of rice varieties, where co-farmers predominated. Government technicians and co-farmers shared the function of creating awareness of crop protection measures among the farmers in Village 2. In Village 3 the communication network for first information sources was comparatively varied, consisting of government technicians, commercial personnel and co-farmers. Village 4 was similar to Village 3, although co-farmers figured prominently as first information sources for rice varieties. Commercial personnel held sway for information on control of insects, weeds and rats.

For additional information the Villages relied mainly on government technicians although dependence on co-farmers was also noted, especially for rat control and selection of rice varieties. In addition, the farmers relied on their own experience, especially for weed control. In Village 3, government technicians retained their significant role as additional information dispensers, especially for plant disease and rat control. However, commercial personnel emerged as important sources of instructional information, without the co-farmers relinquishing this particular communication role. Village 4 exhibited a more intricate communication network where additional information generation was shared among government technicians, commercial personnel, co-farmers and self-experience. It seems therefore that, as villages become more accessible, the communication becomes less dependent on a single information source and more dispersed.

In Village 1 the legitimation sources were mainly government technicians, particularly for insect control. However, there were many cases of self legitimation especially in the area of rice varieties and weed control. In Village 2, again government technicians led as legitimators, except in the selection of rice varieties, where co-farmers and self-experience were mentioned more frequently. The legitimation pattern in Village 3 relied heavily on government technicians and commercial personnel and to some extent co-farmers. Village 4 had a relatively extended pattern, consisting of government technicians, commercial personnel, co-farmers and self-experience, with varietal selection increasingly becoming a function of co-farmers and self-experience.

For all villages in the study decision making was largely the function of the farmer himself, in some cases involving government technicians as decision makers. This pattern was especially marked in Villages 1 and 2. In Village 4 some variation was noted in that commercial personnel and co-farmers also served as decision makers.

26
Conclusions And Further Recommendations For Research And Development

E.J. Tait

The papers in this volume are not, in any sense, the end of the program of research and development identified in 1979 by the Perception and Management of Pests and Pesticides (PMPP) network. However, this is an appropriate time to reappraise research directions and to take stock of the implications of research already done, based on discussions at the Chiang Mai meeting and on other occasions, correspondence with network members and changes in national and international circumstances.

RESEARCH OUTCOMES AND THEIR IMPLEMENTATION

The papers presented in this volume enable one to make comparisons between pest and pesticide management in developed and developing countries, between technologically advanced and subsistence farmers, between perceptions and decisions taken at the national or regional level and those at the community level, and among a wide range of sources of pressure and influence on decision makers. A detailed analysis of this nature could occupy another entire volume. Here there is only space to give a few examples, leaving the rest to the interested reader.

As would be expected, there are dramatic differences in the management of pests and pesticides between developed and developing countries. For example, developed countries are generally struggling to cope with food surpluses, produced by only a small proportion of the total population, a success story partly attributable to the use of pesticides. On the other hand, many developing countries have a serious food deficit, but have a majority of the total population engaged in agriculture. Thus, in the papers by Dearden and Carr, pressure is brought to bear on the use of pesticides for amenity and agricultural purposes, by a largely non-agricultural population, often because of fears of environmental side effects. Conversely, in developing countries, as described in most of the Part II papers, pressures, either for or against the use of pesticides, stem largely from the agricultural community itself, or from government or industry sources — there is little or no reference to a generalized public concern about this issue. Where pressure groups and non-government organizations are active in developing countries, as described in the paper by Mohan, the major perceived problems are the health of agricultural and factory workers and the development of pest resistance.

Within developing countries there are marked variations in the crop protection practices of farmers at different levels of technological evolution. The papers by Goldman, Tukahirwa and Samonte et al. each make comparisons between groups of farmers at progressive stages of development. Subsistence farmers, as described particularly in the papers of Youdeowei, Tukahirwa and Conelly, are often unable to obtain the pesticides they need to grow adequate crops. Farmers who are beginning to make use of technological inputs, as described by Goldman, Hussein, Heong and Ho and Samonte et al. are often struggling to come to terms with the demands of the new technology, with only sporadic and inadequate advice that is rarely tailored to their real needs. At higher levels of sophistication, farmers may have reached a point where their pesticide use is excessive, leading to concern about its sustainability, and attempts are being made to help them to adopt more rational methods of crop protection, as described by Kenmore et al., Medina, Fagoonee and Waibel.

In Part 1, Black et al., Evenson, Zaidi and Atteh describe attempts to improve the rationality of pest management practices, from the perspective of government or other official agencies. The varying degrees of success in implementation of these projects can be related, among other things, to the extent to which those in authority are aware of the needs and motivation of farmers, and of the constraints operating on them. The same is true of public health programs that involve local participation, as described by Mouchet and Guillet.

In discussions during the conference, the wisdom of some government attempts to promote the use of pesticides was questioned. Such schemes are certainly justified in pest or disease outbreak situations where regionally co-ordinated action may be needed, and where poor farmers suffer regular crop losses and cannot afford the necessary pesticides. However, under normal circumstances, government efforts may be better spent on developing IPM programs and devising other ways to improve the rationality of pesticide use.

As they adopt increasingly technologically-based pest management systems, farmers often lose control of crop protection on their farms. They either spray on an insurance basis, regardless of need, or they spray on the recommendations of a government or commercial adviser. There may be occasions when this is the best option in the short term, but it has potential dangers in the longer term. The papers by Kenmore et al. and Lane and Tait deal with the need to train farmers to recognize pests, diseases and weeds on their farms and to decide for themselves on the need to treat with pesticides.

One point emerging clearly from many of the papers is that the acceptability of the pesticides used, and of the extent of their use, depend very much on the perceptions of the people concerned and, even within a single cropping system, there may be no consensus of opinion. Research can elucidate the nature and extent of such disagreements, but it cannot change the political nature of the final decision, whether it be taken at the farm, community or government level. Where a particular behavior pattern has been identified as undesirable, it is often surprisingly resistant to change by the straightforward provision of information and training. Knowledge of the underlying perceptions and motivation is necessary for the design of effective means of influencing behavior.

A strong thread connecting several of the Part I papers is the need for better monitoring of the operation of pesticide control systems, which often fall short of official expectations or assumptions. In promoting the use of pesticides, government bodies may adopt a variety of methods with little monitoring of the expected outcome and no learning from mistakes made or problems encountered. Attempts to control the distribution and sale of pesticides also may not live up to expectations (see the papers by Ahmad and Atteh). The papers by Lim and Ong, Tukahirwa, and Sharma deal with the monitoring of actual and potential environmental side effects of pesticides. The nature and extent of toxic reactions in the human population are discussed by Mohan, Atteh, Sharma, Medina and Fagoonee. The need for better monitoring of pesticide production and use and the biases incorporated in different statistics are discussed by Tait and Lane.

A related point emerging from the conference discussions was the importance, in developed countries, of national and international non-government organizations and interest groups, that act as watch-dogs on behalf of the public. When dealing with complex technology, such as pesticides, their role in influencing government policy was seen to have been a valuable one and their emerging influence in developing countries was seen as a hopeful sign which should be encouraged.

The authors of many of these papers are actively engaged in the implementation of pest management systems on farms, particularly IPM, and can write from experience. Since the PMPP network was set up, the importance of such community-level research has become much more widely appreciated and, as these papers show, progress is being made in developing appropriate methods. Researchers are also increasingly aware of the obligation to press home their research findings, to ensure that they are taken account of by those in a position to influence events, rather than seeing publication of the findings as the desired end-point.

Implementation of research at the regional, national and international levels, usually has to take place via the policy making process, requiring interpretation of the results within a policy analysis framework. Awareness of this need, and an understanding of how it can be met, has been slower to develop, being currently about ten years behind that of community level research in the pest and pesticide management areas.

FUTURE RESEARCH AND DEVELOPMENT NEEDS

The major issues identified as urgently in need of further research and development work are outlined below.

The Operation of Regulatory Systems. Progress in the regulation of pesticides is being made at government and organizational levels. However, there is anecdotal evidence of the failure of many systems to operate effectively in practice, in both developed and developing countries. Policy studies are required to indicate the nature and extent of such short falls and suggest remedies. Comparative analysis of the experience of different countries could help in the design of successful regulatory strategies, particularly for small or poor countries that cannot afford to reproduce a comprehensive regulatory system like that of the United States

Agrochemical Industry Studies. More research is needed on the agrochemical industry itself, on decision making on the production and marketing of pesticides and the development of new chemicals, and on the role of the industry in influencing farming and public health pest control systems. The industry is an important source of pest control expertise, and at the very least its activities cannot be ignored. At best they can be integrated with those of government and the local community, but this requires a much more detailed understanding, by one of the other, than usually occurs.

Crop Protection Research at the Farm Level. Much more work remains to be done on the design and implementation of systems to provide effective crop protection with the minimum of side effects on people and the environment. A distinction needs to be made here between (a) farmers who are already using modern pest control technology but who need help to improve its efficiency, and (b) those who could benefit from it but are, for one reason or another, denied access to it. In the first category, a major issue to have emerged from research done under the PMPP program is the gap between the way farmers actually use modern technology and the instructions given to them. Central to this problem is the 'top-down' system of information communication, that assumes infallibility of expertise among scientists and other specialists, tends to ignore what farmers already know, and often fails to make the effort to understand fully the nature of the farmer's problems and needs. The major issue for the second category of farmer is the need to develop appropriate technology for food crop use and to provide access to it.

Public Health Pest Control. There is a need for a more integrated approach to public health pest control, linking it where necessary, to agricultural pest control, and giving greater coordination between measures taken at local and national levels. The work already being done by the World Health Organization in these areas needs to be supplemented by perception-related research that would help to indicate why there is resistance to integrated vector control and to community participation, why there is a lack of coordination between agriculture and public health, and how to improve these situations.

New Crop Protection Developments. New developments in biotechnology and in computer-based aids to decision making could bring about dramatic changes in agricultural and public health pest control. There will be attendant impacts on the operation of the agrochemical industry, regulatory systems, government extension services, farming systems and health care. The relationships between technology and policy at the macro-level, and between perceptions and behavior at the micro-level will be important new research areas.

RESEARCH AND DEVELOPMENT METHODS

The title of the PMPP network reflects its emphasis on the role of perceptions (used here as synonymous with attitude system, value system or world-view) in determining behavior (the relevant behavior being the management of pests or pesticides). Research is more productive when it deals with both sides of this equation, exploring the interactions between perceptions and behavior and noting the circumstances under which the two are, or are not, related to one another.

The emphasis on perceptions is part of a more general concern to foster an holistic, or systemic approach to management problems, as in integrated pest management (IPM). This means having a group of people with skills in all the relevant disciplines working closely together on a problem and doing research as part of a continuing development program. In the case of an IPM program, this should include the active participation of the client farmers. Undertaking such an approach implies the following commitments:

i)   to move from mere cross-disciplinary exchanges to full partnership, learning enough of the partners' disciplines to ask meaningful questions or pose meaningful problems within their terms of reference;

ii) to submerge individual disciplinary interests in favor of problem solving studies directed to practice, and avoiding the temptation to concentrate on safe, publishable exercises;

iii) to maintain a dialogue with the client farmers at all stages of the project work, being careful to avoid the temptation to retreat into reassuring conversations with colleagues;

iv) to appreciate the full diversity of farmers' current practices and their effects on any planned system;

v)   to involve the clients (pest managers) as partners in the design and presentation of training materials.

Much remains to be learned about the study of perceptions in developed and developing countries and about the application of such research within a systemic framework. There are also many difficulties in adhering to the commitments outlined above but they will be outweighed by the rewards in practical application.

CONCLUSIONS

At the close of the PMPP meeting in Chiang Mai, the participants agreed on the following list of problems identified and recommendations. They provide a convenient summary of this chapter and the preceding papers.

Pest Managers and Pesticide Users

The over-use and inappropriate use of pesticides is leading to undesirable effects on agricultural systems, on farm workers applying pesticides and on the environment, and also to exacerbation of some public health problems. Research is needed on the role of the agrochemical industry and government advisory services in encouraging appropriate use of pesticides.

There is a need for multidisciplinary and systems approaches to research and development on crop protection. For effective implementation of integrated pest management programs, all disciplines should work together in a project, including also the active participation of the end-users (farmers). It is important to include in this an appreciation of the diversity of current agricultural practices.

Communication channels at policy, practitioner and research levels should be improved, to communicate multidisciplinary recommendations more effectively and to raise the awareness of funding agencies to the need for a more multidisciplinary approach.

Research and recommended practices should be more adaptive and capable of fine tuning to cope with rapid changes in pest problems and in the socio-economic environment.

PMPP Profiles

Further profiles, describing national arrangements for controlling the production, distribution and use of pesticides in developed and developing countries, are required to identify problems in current practices and suggest necessary changes. There should be a standard framework for these studies to enable comparisons to be made between countries and to encourage the recommendation of policy initiatives.

Active government promotion of pesticide use should only be undertaken after careful consideration of pest management needs, of the likely effects on agro-ecosystems, and of alternative pest management options. Where farmers are actively involved in a market economy, such promotion is less likely to be justified.

Education and Training

There is a need to create a two-way flow of information with a view to developing appropriate and effective means of influencing pest managers, pesticide users and also policy makers at the government level.

In both developed and developing countries, farmers should be encouraged to take responsibility for their own decision making on pest management in close liaison with government advisory agencies.

More appropriate education and training facilities are required to involve local communities in public health projects and to improve understanding of the agromedical implications of various practices.

International Aspects

More accurate information on the international flow of pesticides and the relocation of pesticide production facilities is needed. Comparisons of data from PMPP Profiles of individual countries could go some way to providing this information.

The role of international organizations including the United Nations and non-government organizations, in raising the public awareness of potential problems and making suggestions to governments, should be encouraged and strengthened.

ACKNOWLEDGMENTS

The contents of this chapter are based on discussions and correspondence with participants in the PMPP network, at the Chiang Mai meeting and on other occasions, and are an attempt to summarize their views. In particular, the section on *Research and Development Methods* should be attributed to Peter Kenmore, FAO, Philippines.

Index

Acaricides 32, 188
Acephate 10
Advice *See* Extension
Aedes spp. 110-112, 114
Africa 112, 113, 123. *See also* individual countries
Age 143, 144, 176
Agricultural support policies 10, 117
Agrochemical industry 5, 38-48, 72, 87, 125, 135, 222, 223
  -competition 19, 29
  -control of 73, 80-87, 90, 135, 224
  -market growth 19, 22, 39, 42
  -production statistics 38-48, 222
  -promotion activities 22, 29, 55, 67, 87, 100, 151, 219, 220, 222, 224
  -research and development. *See* Research and development - pesticides
  -salesmen 55, 136, 151, 152, 178, 179, 212-218
Aid. *See* International aid
Alachlor 34, 87
Aldicarb 31, 32
Aldrin 32, 41, 71, 76, 89
Alfalfa 43
Aluminium phosphide 87, 89
Ametryn 34
Amitraz 51, 179, 186
Anopheles spp. 112-114
Antestia 121
Antibiotics 32, 35
Ants 90, 150
Aphids 31, 44, 46, 54, 55, 119, 120, 161, 178, 185
Armyworms 185, 201
Arrowroot 119
Asia, 123. *See also* individual countries
Asulam 34
Atherigona soccata 123, 201
Atrazine 34, 88
Attitudes 2, 38, 44, 45, 52, 53, 55, 102, 107, 132-141, 143, 147, 165, 223
Australia, 49-57
Azinphos methyl 176
Azocyclotin 179

Baboon, as pests 202
Bacillus thuringiensis 32, 35, 50, 51, 153, 155

Household context. *See* Pesticide use-domestic
Hymenia recurvalis 160, 161

-mixed 51, 52, 102, 103
-optimal 66, 130, 199
-rational 28, 45, 50, 103, 147, 175, 220
-traditional 87, 112, 120, 122
Pest outbreaks 15, 17, 18, 22, 66, 74, 76, 89, 100, 120, 150, 192, 220
Pest resurgence 15, 22, 28, 135
Phenazine-5-oxide 35
Phenthoate 10, 32, 153, 169, 179
Pheromones 19, 67
Philippines 13, 98-108, 150-157, 191-197, 210-218
Phorate 17,41
Phosphamidon 87, 179, 185, 187
Phthorimaea operculella 178
Phytophthora infestans 178
Phytotoxicity 28, 125, 145, 159
Pigs, wild, as pests 119, 121, 123, 124, 202
Pineapple 176, 179
Piperophos 212
Pirimicarb 32, 87
Pirimiphos methyl 77
Plant hoppers 15, 76
Ploceus spp., as pests 203
Plutella xylostella 77, 150, 178
Podagrica spp. 160, 161
Podborers 122, 178
Poisoning 11, 12, 15, 17, 31, 41, 73-75, 154-156, 176, 178, 182
Policy 1, 3, 6, 18, 19, 29, 67, 109, 125, 133, 139, 160, 188, 190, 203,
    221, 222, 224
Pollution 76, 184, 186-188
Porcupines, as pests 119
Potatoes 29, 31, 43, 44, 86, 89, 118, 119, 121-124, 127, 151-154, 159, 175,
    176, 179
Poverty 15, 114, 188
Predators, of pests 15, 25, 50
Pressure groups 16, 58-65, 72, 85, 132-141, 219, 221
Productivity 71, 72, 132, 139, 182, 187, 188, 204
Profenofos 53, 54, 125, 153
Profitability, of crop protection 129, 144, 187
Prometryne 88
Propanil 10, 89
Propargite 32
Prophylactic spraying. *See* Pest management strategies - insurance based
Propineb 33
Propoxur 10, 87, 153
Protective clothing 11, 73-75, 90, 152, 180